HOW WE LIVE AND WHY WE DIE

How We Live and Why We Die

the secret lives of cells

LEWIS WOLPERT

faber and faber

First published in 2009
by Faber and Faber Ltd
Bloomsbury House
74–77 Great Russell Street
London WC1B 3DA

Typeset by Faber and Faber Ltd
Printed in England by CPI Mackays, Chatham

A CIP record for this book
is available from the British Library

ISBN 978–0–571–23911–5

2 4 6 8 10 9 7 5 3 1

What am I, Life? A thing of watery salt
Held in cohesion by unresting cells,
Which work they know not why, which never halt,
Myself unwitting where their Master dwells.

– John Masefield

Acknowledgements

I am grateful for the help given me by a number of people. Alison Hawkes read many drafts and corrected both my English and my style. My agent Anne Engel made helpful suggestions, as did Eleanor Lawrence and Helen Jefferson-Brown. Julian Loose at Faber was, as always, invaluable and his co-worker Kate Murray-Browne was an excellent editor.

Contents

1

Introduction

the miraculous cell

Cells are the miracle of evolution. They are miraculous not in a religious sense, but because they are so amazing. Each of us is essentially a society of billions of cells that govern everything, from movement to memory and imagination. We come from just one single cell – the fertilised egg. In fact, all life originated from one cell, billions of years ago.

Arriving at cell theory has been considered even more important to biology than Darwin's theory of evolution, because it brought together so many diverse phenomena under a common concept. It explains how our body functions, which is vitally important when things go wrong and we get ill. We need to understand how the units in this society of cells function to understand a great variety of illnesses – from cancer to strokes to Alzheimer's disease. The future of medicine lies in cell science. We also need to understand why this community of cells ages, and the nature of death itself. When we are infected by disease-causing bacteria and viruses, the immune system tries to identify the foreign invaders and destroy them, and indeed goes to great lengths to do this. What drives and underlies these fundamental processes of life other than cells? And what is life itself?

Only a few years ago, discussions about cell biology were limited to a handful of scientific experts with little contact with the public. Today, they are in the news for both medical and ethical reasons. There are almost daily reports on stem cells, which may hold the promise of curing numerous diseases; on cloning; on increases in cancer and obesity; and on the use of DNA to detect bad genes and identify criminals. Cell biology is now the focus of general interest or alarm. Understanding how cells function helps to clarify these contentious issues.

Cells are the basis of all life, from thousands of different bacteria to the thousands upon thousands of different animals and plants. Cells are very small – there are a million or so skin cells between your toes and your nose – but for their size they are the most complex objects in the universe. Going up the scale, the collection of cells in our brains would see off all rivals for the complexity prize. And yet there is no overall controller of this cellular society: it is a true co-operative.

There is nothing living that is not made up of cells, even though their forms can vary from snails to elephants to roses. Cells can do a remarkable number of different things: skin cells cover and protect us, muscles contract, nerve cells conduct impulses, gut cells absorb food; cells form the vessels of the circulatory system, cells in the kidney filter the blood, immune system cells protect us from foreign invaders, blood cells carry oxygen, bone and cartilage cells give us support; and so on. Yet, in spite of the apparent differences between, for example, a nerve cell and a skin cell, they work by the same basic principles.

Each cell is surrounded by a thin and flexible membrane that controls what can get into and out of the cell. In many cases this membrane is in tight contact with neighbouring cells which are held together to form extensive sheets of cells, such as our skin and blood vessels. Inside each cell there are two major regions: the nucleus and the cytoplasm. A bit like a flattish disc, the nucleus is the most prominent structure and is itself bounded by a membrane that controls what can enter or leave the nucleus. Nuclei contain the DNA code for determining the sequence of amino acids in our proteins, but the proteins are made in the region surrounding the nucleus called the cytoplasm, where most of the cell's functions are performed by proteins in a fluid environment. Also in the cytoplasm are small sausage-shaped units, the mitochondria, which provide energy for the cell, and other groups of membranes. Our red blood cells are an exception: here, all these elements have been thrown out during the cells' development.

That our cells are so similar is hardly surprising, as they all develop from just one cell: the egg. From the fertilised egg many different structures in our body develop, some of astonishing complexity like the brain, and this development involves all sorts of cell activities such as the cells knowing where they are, changing shape to mould the embryo, and dividing to give rise to hundreds of different types of cells, from red blood cells to nerve cells.

Our brain and spinal cord, for example, come from an initially flat sheet of cells that folds up to form a tube. This extraordinary structure has its programme for development in the tiny fertilised egg. The brain itself, and all the rest of the nervous system, are no more than an impossibly complex set of millions of interacting nerve cells. These interactions are based on the electrical impulses that nerve cells send along one of their long extensions. The nerve cells send messages to our muscles to make them contract. They also interact with each other in ways that enable us to learn, think and feel, and in addition there are special cells that are sensitive to light in our eye, others that are sensitive to a variety of smells in our nose, and others in our skin which can feel pain and detect touch and temperature. Emotions – love, sadness, pain – have their origins in cells.

It was cell division that led me into cell biology at the age of twenty-five. I had been trained as a civil engineer and had worked doing research on the mechanical properties of soils in relation to the stability of foundations. But this was not sexy enough for me; I wanted to do research on something more exciting, and many friends knew that I was searching for a new field of enquiry. Very fortunately, one close friend was getting married away from his home town, and he went to the local library while preparations for the party were being made by others. He was already interested in cells and came by chance on a paper in a journal discussing the application of mechanical principles to attempts to understand how a cell divided into two. How did a constriction form on the surface of a spherical cell and so divide it into two cells? He wrote to me and suggested that this was the problem that I should work on. I did, and fell in love with cells and embryos.

My first passion was for cell division in sea-urchin embryos. These are easily obtainable at marine stations in the summer, and their large eggs are suitable for experiments on the mechanical properties of the membrane. Later, I studied the development of the urchin itself, then the regeneration of hydra, and finally the development of the chick limb. Cells seemed to me, and still do, miraculous.

But how much is really understood about the secret life of cells? This book hopes to answer the essential questions about cells, and explain how they function. I will focus on the human body, even though other organisms, from bacteria to worms and flies, have been invaluable as models for studying and understanding human cell behaviour. However clever one thinks cells are, they almost always turn out to exceed one's expectations.

Imagine that you are almost unimaginably small – no bigger than a water molecule, for example, which is very small indeed: there are more molecules of water in a glass of water than glasses of water in all the oceans. You are so small that you can enter unnoticed into one of the billions of cells that make up the human body. As you approach the cell it will seem enormous, and as you get closer you see that it is enclosed by a membrane. You also see small openings in the membrane through which you could pass, and special ones for favoured molecules. There are also exits where, for example, sodium is continuously pumped out.

Entering through one of the openings in the fatty membrane surrounding the cell, you find yourself in the interior. Here there is an extraordinarily crowded world of millions of molecules – proteins, which are the workers in the cell, and also some sugars and fats – all moving around with extraordinary rapidity. There are also many membranes folded together, and filaments and micro-tubules. The filaments and tubules are formed from proteins joining up to make long shapes. It gives you the feeling that by comparison, being in an enormous football crowd is like being alone. It seems like chaos, but it is not.

The thousands of different proteins are rushing about finding out what work they have to do. The proteins, made from amino acids, look like complex twisted chains; they have extraordinary shapes, and some are acting like machines, cutting up some molecules and constructing others. In doing this they change their shape rather like contortionists. You also see some tiny particles being moved across the cell by motor proteins using fine tubules as railway tracks. Of particular importance is protein synthesis, in which messengers from the genes determine that the amino-acid building blocks are joined together in the right order. These are joined together, one at a time, into long chains, which then fold to give the proteins their complex shape.

The energy for much of this activity is provided by the local energy currency, adenosine triphosphate, commonly known as ATP, which floats around waiting to be spent. Lots of ATP is pouring out of large sausage-shaped cylinders: these are the mitochondria where the ATP is made using the energy from burning glucose.

As you move further into the cell you see the membrane around the cell nucleus, which has entrances and exits for special molecules. When you enter the nucleus, you see a set of very long thick fibres — the chromosomes that contain the DNA of the genes. In some regions there is activity: the genes are being read by a protein machine to make messages for constructing other proteins. A gene is passive and does nothing while its code for the protein is being read. As you move along the chromosome you see that there are thousands of genes.

A little bewildered by now, you are beginning to get some idea of just how complex the machinery is of a single cell. But the single cell is the basis of all life, and in each of our bodies we have billions of them. In this book I will attempt to unravel the mystery, power, and above all the sheer cleverness of our society of cells.

We look first at just how this society of cells was discovered.

2

Discovery

how science made plain the facts of life

In ancient times, simple explanations of life and death, and practical rules and systems of right and wrong, were mainly embedded in religious beliefs and passed down through countless generations. They were expressed in many different cultures in the forms of myths and legends, ritual songs and dances, laws and language. Then came the Greeks. They knew nothing of cells but they tried to understand life, particularly when it went wrong and we became ill.

Science has its origins with the ancient Greeks – the only society which attempted to understand the nature of the world in terms of evidence and logic. Other societies, like the Chinese, had excellent technology, but this was not based on science. As far as biology is concerned, the ancient Greeks did not do very well, because common-sense views are in conflict with scientific explanations, and they could not have conceived the idea that life is based on cells. But Aristotle made great progress with logic, Euclid with mathematics, and Archimedes, probably the greatest scientist, with the physics of levers, specific gravity and floating bodies. Some of the ancient

Greeks were atomists and believed the world was constructed from tiny particles, an idea that Aristotle dismissed. He believed that matter was seamless.

Those early Greeks believed that everything was made out of the four elements: earth, fire, water and air. From this belief they developed the theory that the make-up and workings of the human body were based on four substances called humours: black bile, yellow bile, phlegm and blood. The idea that all diseases and illnesses resulted from an excess of or deficit in one of these four humours was adopted by ancient Greek and Roman physicians and philosophers. Hippocrates, who lived in Greece around 400 BC, promoted this theory and was among the first to reject more mystical explanations of the cause of illness.

From Hippocrates onwards, the humour theory remained the most commonly held view of how the human body functioned until the eighteenth century. Essentially, humours were held in balance when a person was healthy. Greeks and Romans and the Western civilisations that adopted classical philosophy believed that each of these humours would wax and wane in the body, depending on diet and activity. When a person had a surplus of one humour, then that person's personality, and eventually health, would be affected. For 2,000 years this led to doctors bleeding patients to rid them of unwanted humours – a practice of zero use, though perhaps they were helped by the placebo effect.

As regards the origin of life, some thought it came from water and some from the air. The Greek philosopher Empedocles thought that plants and animals arose through the action of fire, which cast up shapeless lumps out of the earth's interior, some of which formed into men. All these ideas derived from the notion that life was some sort of vital force that activated living organisms. This fitted well with the later view of the Christian church: God, as the Bible made clear, had created life. Such views hardly changed in the West until the seventeenth century. Without the invention of the microscope, we might still be holding them.

Another ancient view, which came from China and is still current in modern theories of acupuncture, is based on the belief that the body has a circulating energy or life force (*chi* or *ki*) that travels through invisible channels called meridians. According to this theory, all diseases are the result of disruptions in the flow of one's *chi*. Acupuncture points, it is still claimed, are located on the meridians and represent areas where the flow of energy can be altered, and various illnesses cured.

Cells did amazingly well in hiding their secrets.

Around 2,000 years ago, after the invention of glass, some Romans started looking through it and testing its surprising properties. How could it be so solid yet let light through? They experimented with different shapes of clear glass, one of which was thick in the middle and thin at the edges, and they found that if you held one of these 'lenses' over an object, the object would look larger. But these lenses were not much used until the end of the thirteenth century in Italy, when spectacle-makers began producing lenses to be worn as glasses.

The early simple 'microscopes', which were really only magnifying glasses, had one lens which could increase the size of an object nearly tenfold. This was not enough to see a cell, though it was clearly interesting to use it to look at fleas and other tiny insects. But then came a crucial step forward, around the year 1590, when two Dutch spectacle-makers put several lenses in a tube and made an important discovery. The object near the end of the tube appeared to be greatly enlarged, more so than could be achieved by any simple magnifying glass. They had just invented the compound microscope. There was now the possibility of seeing a cell.

The first person to see a cell – though he did not at the time understand what he was looking at – and to give it its name was Robert Hooke in 1665. As a boy, Hooke was an extraordinarily quick learner, and possessed a manual dexterity that enabled him to build an impressive array of mechanical devices. He left Westminster School in London to take up a poor scholar's place at Christ Church, Oxford. He became accomplished in a variety of fields, including astronomy, biology, physics and architecture, and his skill as an instrument-maker gave him a high reputation. He became a member of the Royal Society of London in 1663, and later its president.

In published researches covering nearly forty years Hooke was constantly casting around for a consistent, underlying principle that could be shown to bind the whole of nature together in what he hoped would be a 'Grand Unified Theory'. That nature did contain grand general principles would have been taken as axiomatic by Hooke, who believed that the entire universe was the product of divine intelligence, and it was inconceivable that God could be inconsistent in His grand design. And as human intelligence was related to that of God, it stood to reason that the key should be within man's reach. As

Kepler had said, science was thinking God's thoughts after Him.

Hooke developed his own compound microscope. When he cut thin sections of cork with a very sharp penknife and examined these through his microscope he observed cell walls and the empty spaces they enclosed. He called them cells – from the Latin *cella*, meaning small room – for they reminded him of monks' cells in a monastery. But he had no idea what they really were, or what their function might be. He thought they were channels for the flow of fluid in the plant. Six years after the publication of Hooke's observations, the Royal Society in London received two manuscripts from other scientists that showed that plants were made of cells with thick walls, and this caused a dispute as to who had made the discovery first.

Hooke's *Micrographia,* which contained the first drawing of a cell as well as his remarkable sketches of numerous small creatures, appeared in the bookshops in January 1665. Samuel Pepys was transfixed in his chair until two o'clock in the morning; he called it 'the most ingenious booke that ever I read in my life'. *Micrographia* was one of the formative books of the modern world.

The person credited with first seeing an animal cell is Anton van Leeuwenhoek, the true father of microscopy. He was self-taught and never attended a university, but he became one of the greatest and most expert micro-scopists. In 1648 Leeuwenhoek, then sixteen years old, went to Amsterdam to learn a trade and became an apprentice in a linen-draper's shop where magnifying glasses were used to count the threads in cloth in order to determine their quality. Developing the idea of the glass-

es used by drapers, he constructed his first simple microscope. He seems to have been inspired to take up the
microscopy of living objects by a copy of Hooke's
Micrographia. He taught himself new methods for grind-
ing and polishing tiny lenses of great curvature, which
made an object look nearly 300 times bigger, and using a
microscope he discovered in 1670 what he called animalcules – little animals – in pond water. These included
moving single-celled organisms; however, Leeuwenhoek
had no idea that some of these were cells, only that they
were small and alive. The first ever picture of a bacterium can be seen in a drawing by Leeuwenhoek in the
Philosophical Transactions of the Royal Society for 1683.

The first animal cells to be seen were blood cells. In
1673 Leeuwenhoek took some blood from his own hand
and noted that it consisted of what he called small round
globuls. He came to believe that all animal matter was
composed of what came to be called globules.
Leeuwenhoek's theories about the body being made of
globules like blood cells was not widely accepted, as they
did not seem in any way applicable to solid tissues. More
popular was the idea that fibres were the basic building
blocks of the human body, and there was little, if any,
progress in understanding or exploring cells for the next
hundred years or so.

In the early nineteenth century many scientists were
looking at biological structures, and René Dutrochet in
France in the 1820s became interested in embryos after
studying medicine. Examining the brain of a snail, he saw
some of the cells quite clearly, as they were very large – a
genuine body cell was observed for the very first time,
some 160 years after Hooke's observation. But great

problems remained. Dutrochet observed the moving animalcules in stagnant water, just as they had been described by Leeuwenhoek, and he became particularly interested in the their endless motility. This led him to think that it could well be a direct expression of the famous and elusive 'vital force' and, hence, one of the means of studying it. As early as 1824 Dutrochet wrote explicitly: 'Life, as far as the physical order is concerned, is nothing more than movement; and death is simply the cessation of this movement.' Dutrochet believed that new globules were formed within old ones and did not accept the idea of spontaneous generation.

Another Frenchman, Jean-Francois Raspail, who was at one point a candidate for the French presidency, was convinced that cells were fundamental and, like Dutrochet, believed that new cells were formed inside existing ones. Two of his aphorisms are impressive: 'Give me an organic vesicle endowed with life and I will give you back the whole of the organised world', and '*Omnis cellula e cellula*' – every cell is derived from another cell. That was the discovery of the core of life – great progress, but there was still a lot missing. It was not known *how* new cells formed, even though, in 1832, the Belgian politician and amateur scientist Henri Dumortier observed cell multiplication in a plant by true cell division. When the terminal cells grew longer than their neighbours, a partition wall formed, dividing them into two. It is nice to learn that politicians at that time cared about basic biology.

The view of the cell in the early nineteenth century was essentially globule-based, and it was held that there was nothing of interest inside the globule. Yet Leeuwenhoek back in 1682 had reported seeing a structure inside the blood cells of cod and salmon. This was the nucleus, whose function would have to wait till the twentieth century to be understood. The name 'nucleus' was given to this structure by the Scot Robert Brown in 1831; he was sure that it had an important function, but had no idea what it was.

The development of cell theory was greatly influenced by the German physiologist Theodor Schwann, who was a devout Catholic throughout his life. His early research was directed to the nervous and muscular tissues, and his observations with his microscope led to his discovery of the cellular fatty sheath covering many nerve cells, which bears his name - the Schwann cells. He broke with the traditional, rather mystical view about life forces and worked towards an explanation of life based on physics and chemistry. When he was dining with the botanist Matthias Schleiden in 1837, the conversation turned to the nuclei of plant cells. Schleiden was a lawyer whose hobby was plants, which he studied with a microscope. Schwann remembered having seen similar structures in the plant cells that he had seen in a vertebrate embryo, and they instantly realised the importance of connecting the cells in animals with those in plants. Here was the true beginning of 'cell theory'.

In 1838 Schleiden suggested that every structural element of a plant was composed of cells or their products. The following year a similar conclusion was elaborated for animals by Schwann in his famous *Microscopic Investigations on the Accordance in the Structure and Growth of Plants and Animals* (1839). Here he stated that 'the elementary parts of all tissues are formed of cells'. The conclusions of Schleiden and Schwann are considered to represent the official formulation of cell theory and their names are almost as closely linked to cell theory as are those of Watson and Crick with the structure of DNA more than a hundred years later, for which they deserve enormous credit. But Schleiden and Schwann made terrible errors in their attempts to understand how cells reproduced.

It is worthwhile thinking about the problem facing them. If you had a microscope, where would you look to see how cells were formed? There is no obvious place to look in any easily available animal. A very good place to look would be the developing embryo, but you would first have to know that all the cells in the body come from a single cell, the egg, by cell division. Schwann and Schleiden did not know that; it was not recognised until later. According to Schleiden, however, the first phase in the generation of cells was the formation of a nucleus of crystallisation in the space outside the cells, with subsequent progressive enlargement of such condensed material to become a new cell. Schwann also believed that the formation of new cells involved a kind of crystallisation, beginning with the deposit of a crystal around a granule in the space between cells.

They did not know about, or ignored, the observation by Dumortier in 1832 that the terminal cells of the plant *Conferva*, a green alga with branching filaments, multiplied by cell division. For animal cells, the key observation was that made by the German scientist Robert Remak in 1841 on dividing chick embryo cells. He was critical of the Schwann and Schleiden model. Another of Remak's key observations was that the frog egg underwent a series of divisions to give rise to the cells of the embryo. In addition he said that tumours came from the division of normal cells.

Writing in 1859, the German doctor Rudolph Virchow proposed a very important concept of life: 'Every animal is a sum of vital units, each of which possess the characteristics of life . . . It follows that the composition of the major organism, the so-called individual, must be likened to the kind of social arrangement or society, in which a number of separate existences are dependent upon one another, in such a way however, that each element possesses its own particular activity, and although receiving the stimulus to activity from other elements carries out its task by its own powers.' Here we have for the first time the concept that we are a society of cells. He also emphasised that cells came from other cells, and his 1858 aphorism *omnis cellula e cellula*' (every cell from a pre-existing cell), which he may have stolen from Raspail, became the basis of the theory of tissue formation, even if the mechanisms of new cell formation were not yet well understood. This was made even clearer when the development of the frog egg was observed and the division of the egg was seen to give rise to many cells, some of which developed into muscle and cartilage cells.

It was not easy, however, to recognise that nerves too were made up of single nerve cells, as their longest extensions – the axons – could be more than a metre long, and they bore little resemblance to the common small rounded cells described elsewhere in animals. Attempts at reconstructing a three-dimensional structure of the nervous system were frustrated by the impossibility at that time of determining the exact relationships between nerve cell bodies and their many long extensions.

The important breakthrough came in 1873 when an Italian doctor, Camillo Golgi, developed a technique for staining nerve cells that made them black: even the blind, he said, could now see a full view of a single nerve cell and its extensions, which could be followed and analysed. He believed that the branched axons stained by his black reaction formed a gigantic continuous network along which the nervous impulse propagated. In fact, he was misled by an illusory network created by the super-imposition and the interlocking of axons of completely separate cells.

Matters changed quickly in the second half of the 1880s. In October 1886 the Swiss embryologist Wilhelm His put forward the idea that the nerve-cell body and its extensions formed an independent unit. The nervous system began to be considered, like any other tissue, as a sum of anatomically and functionally independent cells, which interact by contact rather than by continuity. Definitive proof of the nerve-cell theory was obtained only after the introduction of the electron microscope in the 1930s, which allowed identification of the junctions – synapses – that exist between nerve cells.

The structure of cells now became an increasingly important issue. The nucleus in animal and plant cells was better defined, though no one knew then that it contained genes. The road to this understanding required a better description of cell division. It was observed in both plants and animals that at cell division the nuclear membrane dissolved and a spindle-shaped structure appeared across the cell. Some granular material collected in the centre of the cell on this spindle, and then moved to each of the future daughter cells. There was then a constriction that divided the cell into two, and the nuclei reformed in the two daughter cells.

That material that collected on the spindle contained a preliminary clue to our heredity – but who could have realised that! In the 1870s it was clear that the material was made up of little rods of different sizes and that each of them divided into two, and then one of each pair was moved along the spindle to the future daughter cells. These rods were called chromosomes. Edouard van Beneden, who was a professor of zoology in Leiden, found when studying a parasitic worm that it had very few chromosomes, some species having only two. He could thus easily follow the behaviour of the chromosomes at cell division, or mitosis as it came to be called. He confirmed that the chromosomes split longitudinally and one of the pair went to each daughter cell. In the process he had discovered a key feature of how sexual differences are specified.

In sexually reproducing animals like we humans, half of our chromosomes come from the mother and half from the father. This occurs at fertilisation; in humans each parent contributes 23 chromosomes, so we have 46 in the fertilised egg and our body cells. Van Beneden was studying an animal that has only two chromosomes in each cell, and realised that one of these was from the sperm and the other from the egg. Thus egg and sperm have only one chromosome each, and when the sperm fertilises the egg the egg has two chromosomes, and it gives rise to all the cells in the animal. But how then does the adult with two chromosomes in each cell produce sperm and egg with only one chromosome? It turned out that this involves a halving of the number of chromosomes, known as meiosis, when the egg and sperm develop. They go through two cell divisions but only double the number of chromosomes once, thus halving the number of chromosomes. Meiosis is a fundamental feature of genetics, and in humans reduces the number of chromosomes from 46 in adult cells to 23 in human sperm and egg.

A central unsolved problem of the time was how characters were passed from parents to offspring. This was an especial problem for Charles Darwin with his theory of evolution by natural selection (first published in 1859). How did the variation on which selection acted arise? Darwin, who did not know about cells, believed that the material from which the embryo formed budded off from all parts of the parents' bodies. This became known as pangenesis. Darwin's cousin Francis Galton distrusted pangenesis and proposed that special material was somehow transmitted from one generation to the next.

A fundamental insight into genetics was provided by Gregor Mendel. The story is complex, for not only was his discovery neglected, but his own ideas about its significance are far from clear. Mendel's experiments on peas in the monastery garden at Brno were published in the Brno natural history journal in 1866, but attracted very little attention. He had shown that the characters of the seeds of peas – round or angular, yellow or green – could be understood as features that were inherited almost as 'particles'. But some features were not always expressed, as they could be masked by a dominant character. When tall and short plants were crossed, the offspring were all tall, and when these were crossed, one third were short, the others tall. There was as yet no talk of genes.

Theodor Boveri from the University of Wurzburg extended van Beneden's observations on chromosomes and showed that their structure was preserved during development. He concluded that they were independent entities. He found that if more than one sperm entered the egg of the sea urchin, the fertilised egg ended up with an excessive number of chromosomes and the embryo was abnormal. Examination of chromosomes in worms and sea urchins led Boveri to conclude that the chromosomes were passed from cell to new cell at division, and that they might be the carriers of heredity. In 1902 Boveri was quick to see the correspondence between his belief in the individuality of chromosomes and Mendel's factors that controlled inheritance.

Boveri was dealing with genes, although they were not called that yet. The term gene had a weird origin. Hugo de Vries coined the term 'pangene' for the smallest particle of heredity in 1889, and it was then abbreviated to 'gene' by Wilhelm Johannsen. In 1908 the Cambridge zoologist William Bateson wrote about inheritance and variation: 'Such a word is badly wanted and if it were necessary to coin one, Genetics might do.'

It was the American Thomas Hunt Morgan who really established the existence of the gene during the first quarter of the twentieth century. Morgan was working as a developmental biologist but was making little progress. He began to study the fruitfly *Drosophila melanogaster* and noticed an abnormal white-eye character (normal fruitfly eyes are red) in one of the flies, which he then showed was linked to the sex of the fly. He then found three more sex-linked characters and was able to show that a particular bit of a chromosome, a gene, was responsible for each character. Genes were clearly located on the chromosomes, and the fundamentals of genetics now became established. Chromosomes were the only way in which characters could be transmitted from parents to their children. The fruitfly remains an important model system for both genetics and development, and Morgan was awarded a Nobel Prize in 1933.

The nature of the gene was still unknown. Most workers thought it was made of protein. The discovery in the 1920s of viruses that could replicate in bacteria confirmed this view, as they were found to be 90 per cent protein. Some evidence that DNA is the hereditary material was discovered in 1928. Fred Griffiths, a British medical officer and geneticist, was studying the bacterium that causes pneumonia and found that there were two forms: a virulent one that caused disease when injected into mice, and one that was harmless. If he heated the virulent strain, it too became harmless. However he was greatly surprised to find that if he injected the heated virulent strain together with the harmless strain, pneumonia developed. Somehow, the heated virulent strain had passed on its virulence to the harmless strain.

It took several other scientists some fifteen years to identify the factor that had been transferred – it was DNA. This was a surprise, as in those days they thought such a factor would be a protein. The discovery that DNA is the hereditary material came in 1944 from experiments with a virus that infected bacteria. The virus contained only DNA and protein, and it was shown that it was only the DNA that contained the information to make more viruses. Later in the 1940s it was shown that bacteria could have their character changed by an agent that was not a protein, but a nucleic acid – DNA. Some were still sceptical that genes were made of DNA, but others were not. The gradually strengthening notion that the hereditary material was DNA and not protein was eventually confirmed in the early 1950s by Crick and Watson's determination of the double-helix structure of DNA, followed quite shortly by the discovery of the mechanism by which DNA duplicated itself and coded for proteins. The molecular basis of cellular mechanisms was beginning to be worked out.

The chromosomes had gained a function, but it was not until the 1950s that the function of mitochondria was uncovered. These small structures had been observed in all cells other than bacteria in the nineteenth century, but their function was not known. The name of this structure was coined to reflect the way they looked to the first scientists to see them, stemming from the Greek words for 'thread' and 'granule'. For many years after their discovery, mitochondria were commonly, and wrongly, believed to transmit hereditary information. In the 1950s they were isolated from cells and their structure was established by electron microscopy. By breaking open a large number of cells and then spinning the resulting mixture of cellular components in a centrifuge, mitochondria, which have a density different from the other structures in the cell, can be isolated. Biochemical studies with isolated mitochondria then identified them as the powerhouse of the cell, where the energy for its many activities, in the form of ATP, is produced.

When doing research on cells it is often necessary to

have access to a large number of similar cells for bio-chemical analysis and, blood apart, this is hard to get from a living organism. A key process for studying cells is the ability to grow them in a culture dish. These techniques were started in the 1890s and a major advance made in 1907 was the ability to culture fragments of tissue for weeks at a time. Cells now had a reasonable life of their own outside the body. Most cells isolated and placed in culture will only grow for a limited period, for reasons that will be discussed later. However, there are some 'immortal' cell lines that have been very useful for studies. One such line, the so-called HeLa line of cells, comes from a woman, Henrietta Lack, who was treated for cervical cancer in 1951. A sample of her tissue was sent to an expert who successfully cultured these cells and established the HeLa cell line. Henrietta Lack was killed by her tumour, whose cells spread throughout her body. HeLa cells are truly vigorous cells and have contaminated many other cell cultures. They have been essential for studying many aspects of cell behaviour, such as the growth of polio viruses in cells, and some 600,000 cultures were shipped to various labs over a period of two years. An important advance was to separate the cells from a tissue into their different types. Even if initial numbers are small, the isolated cells can then be increased by placing them in culture. Most cells placed in a culture dish with suitable nutrients will keep their character: nerve cells will extend their long processes, embryonic muscle cells will spontaneously contract, and cells taken from sheets of cells can form sheets again.

Cell theory was thus definitely established. As well as being the fundamental unit of life, the cell was also seen as the basic element in disease processes, which were due to an alteration of cells' behaviour in the organism. This allowed modern illnesses to be analysed on a cellular basis, as in pathology, which examines tissues with a microscope to find abnormalities that might cause a particular condition. Cell theory stimulated a reductionist approach to biological problems and became the most general structural paradigm in biology. It emphasised the concept of the unity of life and brought about the concept of organisms as 'republics of living elementary units' – a society of cells.

All cells are derived from a common ancestor and they have through evolution conserved their basic common properties. This has the great advantage that understanding and knowledge gained from one organism can contribute, in many cases, to the understanding of other organisms, including ourselves. Humans are not that good a subject on which to do experiments, whereas other species such as mice, frogs, flies, sea urchins, worms and bacteria provide excellent opportunities. The bacterium *Escherichia coli*, for example, which lives in our guts, is easy to grow, reproduces rapidly, and makes about 4,300 different proteins: much of our knowledge of molecular mechanisms such as DNA replication and protein synthesis has come from a study of *E. coli*. Multicellular animals such as the fruitfly have provided the basis of classical genetics as well as the molecular basis of embryonic development. Frogs, mice and chickens have been essential for understanding embryonic development in vertebrates. But here, as we turn to the processes that characterise life, we will focus mainly on human cells.

2

How We Live

how cells replicate, maintain order, evolve and die

It is only from cells that we can find out what life is. Without attempting anything so grand as a comprehensive definition of life, I will describe and examine its key characteristics. The first is the ability to replicate, to create more of itself, as when a cell grows and divides into two new cells; the second is the ability to maintain order and create energy for all the activities in the cell such as movement and the synthesis of molecules; the third is the cell's ability to evolve; and the fourth and last is death. In all of these it is still surprising and unexpected that the key players are the string-like molecules, DNA and proteins.

Proteins are the true wizard machines of the cell. Somewhat less interesting, but of the greatest importance, is DNA, from which the genes in the chromosome are made, and which, as we will see, does nothing but provide the code for making proteins. However, DNA has one fundamental unique feature: it is the only molecule in the cell that is replicated. An identical copy of itself is made before a cell divides. By determining the nature of all the different proteins in the cell, it can also effectively control many cell activities. The DNA in our cells is contained in the nucleus, a flattened disc surrounded by a membrane.

The membrane around a cell is very thin and can be seen only as a black line with an electron microscope, but it defines the border of the cell and separates its contents from the environment surrounding it. This membrane is more than just a barrier, as it contains special proteins that enable some molecules to move quite easily into the cell and out again, while preventing many others from doing so. It plays an active role with its proteins by transporting certain molecules and atoms in and out of the cell. Food has to get in, and waste products out. A special highly evolved feature of the cell membrane is its ability to conduct electrical impulses, the basis of nerve cell function.

Proteins, of which there are around 100,000 different types in our body, are strings of simple small molecules called amino acids linked together. There are some twenty different kinds of amino acid in proteins – our cells can make ten of them, the others must come from food. What makes proteins different from each other is the order and number of these amino acids, as that determines how the protein string will fold, and thus its three-dimensional structure and its function. Proteins are contortionists, folding and changing shape in complex ways. Most of our cells contain several thousand different proteins, and there may be thousands, even millions of copies of each one. If one could enter the miniature society of proteins one would be amazed that anything positive could come from all this activity, for not only are proteins changing shape, but most of them move randomly and rapidly, and come into contact with millions of other molecules every few seconds. Almost every action in the cell is due to proteins interacting with each other or with other molecules such as nucleic acids, carbohydrates and fats.

Of all cell functions, making new cells by increasing in size and then dividing is the most important. New cells come from existing cells by growth, and then division into two. But before we come to the actual process of division we must first look at the preparations for this event. The contents of the growing cell must be duplicated so that each daughter cell after division receives all it requires: genes, proteins, mitochondria, membranes, and lots of other molecules.

All of the structures and molecules have to be divided between the two daughter cells. There thus has to be much growth, and the cell typically doubles in size prior to division. And inside the cell, the DNA in the nucleus must be replicated – it is the only molecule to undergo replication and this is fundamental to life as it provides the information for making proteins which carry out almost all the key processes in the cell. There are about two metres of DNA molecules in our cells and this has to fit happily into the small space in the cell nucleus. Special proteins package these long DNA molecules into chromosomes that fold the DNA into a series of loops and coils so that the DNA does not become entangled with either parts of itself or with other chromosomes. A copy of each of the 46 chromosomes in every one of our body cells has to be passed on to each daughter cell at mitosis.

All the other components of the cell must double up when cells grow before division. Each mitochondrion, of which there are several hundred in most of our cells, will also divide into two and replicate their small amount of DNA. There is synthesis of proteins, fats and sugars, and the proteins play a key role in these processes.

In a small single cell like yeast the process from one division to the next, the cell cycle, is just two hours; most of our cells, if dividing, have a cycle time of around 24 hours. The cycle has four phases. The first and longest is growth; in the second phase chromosomes with their DNA and genes are being duplicated; then there is a third phase, and finally a fourth phase, division by mitosis. As with a washing machine, the cycle is programmed so that the preparations and end result are done in the right order: entry to the third phase before mitosis must not occur until all the DNA has been replicated.

Because of its fundamental importance, it will come as no surprise that there is, as we shall see, a control system for cell division to ensure that actual physical division takes place at the right time, and not before, for example, the chromosomes have been fully and reliably duplicated. Mitosis itself, the division into two cells, takes only a short time, about an hour, but preparation for mitosis can take twelve to 24 hours. There are checkpoints in the cell cycle to ensure that all is going well and, if not, the next step will be delayed until all is in order.

Timing of the sequence of events in the cycle is largely controlled by special proteins called cyclins whose concentration goes up and down during the cycle. They turn off, or on, the function of other specific proteins, and one of the main ways this is done is by either adding phosphorus from the energy molecule ATP to the protein – known as phosphorylation – or removing it. For example, one cyclin starts being made immediately after the cell has divided at mitosis and its concentration increases steadily throughout the cycle until the next mitosis, which it is involved in initiating. Its concentration falls rapidly near the end of mitosis and then starts up again and the sequence is repeated. Other cyclins trigger other stages, such as the initiation of DNA synthesis.

There are also built-in checkpoints. These include an important one near the start of the cycle, before the cell begins synthesising a new set of chromosomes and so doubling its DNA. For example, if there is some DNA damage, then DNA synthesis does not begin. This control involves a very important protein, known as p53, which delays DNA synthesis until the damage is repaired, causing withdrawal from the cycle, or blocking replication permanently by causing the cell to commit suicide and die rather than becoming cancerous because of its damaged DNA. This early checkpoint also makes sure that the environment is favourable for cell replication, by checking whether the right signals and nutrients are present. If the checks are not satisfactory the cell will delay further progress into the cell cycle, and may remain passive for a long time. There is another checkpoint before mitosis, when the cell ensures that all the DNA has been replicated. There are checkpoints that look for broken chromosomes before their separation; if one of these is activated, mitosis is delayed until a repair is done.

The way the chromosomes replicate is based on the special structure of DNA. The key discovery, referred to earlier, of Watson and Crick in 1953 was to show that DNA is composed of two chains of four different nucleotides wrapped around each other in a helical manner. DNA molecules are the key structures of our 46 chromosomes and each chromosome contains a very long DNA molecule composed of two strands wrapped around each other in the famous double helix. Each DNA strand is a string of four different sorts of quite simple chemical sub-units, called nucleotides – that is why DNA is a nucleic acid. The four units are adenine, cytosine, thymine and guanine. In gene language, A, C, T, and G are the letters used to describe the sequence of nucleotides, sometimes called bases, in the DNA which codes for the amino acid sequence of proteins. The bases in DNA fit together in such a way that G in one strand always binds to its partner C in the opposite strand, and T binds to A, and so the double helix is formed. Because one strand of a DNA molecule has a precisely complementary sequence to the other, a DNA double helix can be made into two identical copies of itself by separating its two strands and then making a new complementary strand on each. A single strand acts as the template for its complementary strand. Starting in a region in which the two strands have been separated, a nucleotide is placed on its partner, such as G on to C, by a special protein machine. Then the next nucleotide is given its partner, and so on down the strand until a double helix is again formed. This is what makes DNA unique, and enables it to transmit essential information from one generation of cells – or people – to the next.

Proteins initiate the replication of DNA by prying the two strands apart, starting at many sites. The process is rapid, separating 100 nucleotides per second. Then a protein replication machine synthesises new DNA using the two separated strands as a template and putting nucleotides on to their correct partners. The synthesis proceeds in two different directions and the new strands each form a new double helix. Working all the way down the two strands, the result is two new molecules of DNA, two new identical – providing there have been no mistakes – chromosomes.

When replication reaches the ends of the DNA there are some problems, as there is no room for some of the protein replication machinery on the DNA. This problem is resolved by having a set of repetitive nucleotides at the end that are known as telomeres. These get shorter at each division, unless a special enzyme restores their length. The more a cell divides the shorter the telomeres become, and if they are eventually lost DNA replication is no longer possible. This can be a cause of ageing.

Replication of the chromosomes must be done with extreme accuracy: every nucleotide must be placed with its correct partner only, and if not can lead to a mutation. Mistakes in replication are rare and have been compared to one letter error in copying a thousand books. But errors can result in a mutation with serious consequences if just one nucleotide pair in the chain is altered. There is even a mechanism to try to prevent or correct such mistakes during DNA replication.

Each of the daughter cells must receive a set of genes identical to that of the parent cell. After the chromosomes have duplicated themselves, the two copies of each chromosome remain tightly bound together. They now change their overall form from a long string of DNA and proteins to a condensed form, and become clearly visible under the microscope as solid but thin, threadlike structures. It was in this form that chromosomes were first identified. The cell has then to separate them and move them apart to the future daughter cells, so that each daughter cell has an identical set of chromosomes.

Mitosis is the name for this process, and the key structure is the mitotic spindle. As its name implies it has a spindle-like shape, and is made essentially of microtubules – fine, long tubes assembled from protein subunits. Microtubules have the ability to rapidly disassemble and then assemble elsewhere depending on the local signals. Their relative instability makes them capable of rapidly remodelling, and they are involved in the division of the cell and the separation of the chromosomes. Their basic form resembles a Christmas cracker, with the two ends being star-shaped and called asters. At the equator the middle bulges out, and this is where the chromosomes will move. The spindle is organised by two special centres, protein complexes. At the start of mitosis these two centres sit together on the nucleus. Then, as mitosis begins, they separate and move to opposite sides of the nucleus, where they send out microtubules in all directions that form the star-like asters at each end, and the spindle forms between them. The membrane around the nucleus now breaks down and the chromosomes move to the centre of the spindle. The chromosomes are now attached to microtubules at a special site, with each pair of chromosomes now back to back and so pointing in opposite directions, each to one of the asters at the poles of the spindle. The microtubules interact with the microtubules from the aster opposite them, and the two asters are pushed apart to opposite poles of the cell, the spindle getting longer. The chromosomes now lie on the equator of the spindle, midway between the two asters.

Suddenly, the adhesive force holding the duplicated chromosomes together disappears, and the chromosomes are pulled along the microtubules of the spindle towards the poles at the centre of each aster. Because the attachment sites to the chromosomes face in opposite directions, the chromosomes of a pair are moved to opposite poles, so that an identical set of chromosomes ends up at each pole. It is rather like two individuals standing back to back on a railway line, each with a ring on their belly; a rope attaches to the ring and they are then pulled apart. A nuclear membrane now begins to assemble around each set of chromosomes, forming two new nuclei, and the cell is ready to physically divide in two.

Imagine tying a piece of string around a balloon and then pulling the string so that it constricts the balloon into two smaller balloons. In the dividing cell a contractile ring, which forms just below the cell surface and is positioned between the two nuclei, acts like that piece of string. The position of the contractile ring is determined by the asters. The entire cell membrane of the cell undergoes an increase in tension; because of the asters' interaction with the membrane at the poles, the contraction relaxes there, but continues in the equator where a contractile furrow constricts the cell, dividing it into two. The two daughter cells now separate and start their own lives. The cell has replicated.

The second basic characteristic of life is the maintaining of order within the cell, which involves growth and synthesis of new molecules and the provision of the energy for these processes. The cell is in some ways a tiny chemical factory in which a large number of chemical reactions take place involving small molecules like sugars and lipids, and the making of large molecules like proteins and nucleic acids. Enzymes are very important, as we shall see, as they can make certain reactions possible, and there are many enzyme-based reactions connected in a long series that can lead to the synthesis or breakdown of molecules.

Maintaining order and preventing breakdown into chaos and decay requires energy. There is a universal tendency for all things to become disordered, as clearly stated by the second law of thermodynamics. This law makes clear that the way matter behaves is always to increase its disorder; another way of stating this is that systems move to that state which has the greatest probability. For example, when a handful of coins thrown into the air falls to the ground, there is a tendency for the heads and tails to be disordered, with variations in the number of heads and tails facing upwards on landing. That they should all be either heads or tails is very unlikely. All systems tend to randomness unless energy is used to prevent this.

Synthesis of new molecules as the cell grows, cell movement and muscle contraction, pumping out of sodium – all these activities consume energy, and how cells get their energy is a key issue. Like us, they have to eat in order to survive, and most of their food is used to get energy, though some of it is used as building blocks for new proteins and other molecules. Their food, which comes from what we eat, is their basic fuel. Energy in cells comes from the chemical energy stored in the chemical bonds which hold together the atoms like oxygen and hydrogen in molecules of sugars and fats and other foods.

Animal cells like ours generate energy from the break-down of their food when combined with oxygen, while plants make use of sunlight. The factory for producing energy in animal cells comprises special structures known as mitochondria. These produce sources of energy, mainly the molecule ATP (adenosine triphosphate). ATP is the energy currency of the cell, and the cost of any activity often depends on how many ATP molecules are used. We eat and breathe mainly so that cells can make ATP to provide us with the energy for all our activities. When we take exercise we use up this energy faster, which is why we become tired.

ATP is amazingly versatile as a source of energy for cellular processes, from muscle contraction to protein synthesis. When it provides energy by giving up one of its three phosphate groups it is converted to ADP (adenosine diphosphate). The ADP can then be made into a carrier of energy again by the mitochondria adding back the phosphate group when food like sugars are broken down. There are some thousand million molecules of ATP in a typical cell and all are used and replaced every two minutes. Why ATP was selected in evolution to be a key source of usable energy is not understood, but it works wonderfully well and that is what evolution cares about.

Cells have to take care that when they break down food into molecules for energy or building blocks that they do not break down some of their own molecules by mistake. This is partly avoided by the initial breakdown being in little vesicles called lysosomes, which contain digestive enzymes. Then comes the breakdown of the sugar molecules and the production of some ATP. The next and major stage takes place in the mitochondria, where some

36

of the breakdown products now enter. Oxygen is essential for this step, when most of the energy is produced.

The breaking down by enzymes of a sugar molecule into two molecules contributes to the making of some ATP; there is a coupling of one form of energy, kinetic energy (or the energy from burning glucose), to doing useful work, like making ATP, which is then used to power other reactions. Think of rocks falling off a cliff – the energy of the falling rocks will simply be turned into heat and wasted as they hit the ground. But now put in place a paddle-wheel that the falling rocks hit and turn, and attach by rope to the paddle-wheel a bucket of water which will be lifted off the ground as the wheel turns. The paddle-wheel machinery represents the enzymes that couple these two processes together. Note that the water is a source of energy when it is lifted off the ground: it can be poured into a machine whose function depends on water passing through it like a water wheel – just like ATP, which can be used when required.

Cells do not have our problems with respect to eating, as obesity is not something they have to avoid. Breakdown of food takes place by digestion, mainly by enzymes, outside the cells of our gut. The small units that result from this breakdown can then enter into cells and be used to provide energy. Sugars are a key and even favourite food and they have energy locked in their chemical bonds. When sugars are broken down into water and carbon dioxide – for that is what cells do to the sugars – the energy released is stored in special energy-carrying molecules, particularly ATP. A sugar such as a glucose molecule is cleaved into smaller molecules and for each molecule of glucose that is broken down, two molecules of ATP are made. But other molecules like proteins and fats are also good food. Mitochondria are much more efficient in producing energy and make use of oxygen, and most of our ATP comes from their synthesis. Later we will see that mitochondria are essentially bacteria transformed during evolution to serve our kind of cell.

The production of ATP by mitochondria is a process that has been going on for billions of years and is based on the transport of electrons through membranes, a concept not that easy to understand. There are two stages for the process in the membranes of our mitochondria. A mitochondrion is covered by two membranes, the outer one quite permeable and the inner one much less so. The movement along the membranes of electrons derived from food molecules is used to drive protons – hydrogen ions – across the membranes, and results in an electro-chemical gradient, as the protons carry a positive charge and there are now more on one side of the membrane than the other. The electrons are passed to oxygen, where they form water. Protons now flow back across the membrane down this gradient through a protein machine that makes ATP by adding a phosphate group to ADP. About 30 molecules of ATP are produced for every molecule of glucose broken down.

This is a remarkable mechanism, and its discovery was totally unexpected. Until 1960 it was believed that the way mitochondria produced energy was similar to the way the breakdown of sugar molecules gave rise to some ATP. It was only in 1961 that Peter Mitchell came up with the idea of the proton gradient mechanism, which, not surprisingly, initially met with considerable resistance.

*

The third characteristic of life is evolution, which is the way life and all living organisms from bacteria to plants to ourselves came into being. It is Darwinian evolution to which I am referring. The core of his wonderful discovery is that characters can be passed on from one generation to the next and that this inheritance can change so that organisms can give rise to offspring that differ from the parent. If this gives them an advantage they will be more successful and will increase in numbers compared to organisms that lack that inherited advantage. Basically, evolution is change in inherited characters and selection for advantage. The inheritance of characters is entirely due to genes, which are the only structures in the cell that replicate, and so can be passed from generation to generation.

Differences in the genes and their control regions are the raw material upon which evolution depends, as these determine which proteins are made. The changes must occur in the germ cells that give rise to the next generation, in our case either sperm or eggs. The changes in the DNA are expressed as changes in the nature of proteins as well as where and when they are made. If they help the organism survive they will be retained in future generations; if they make things worse, then the organism with that mutation will soon die out. A mutation can occur within a gene by the alteration of just one nucleotide, which can affect the protein structure.

Finally, a major feature of life is death. This may seem odd, but one must have been alive in order to die. Cell death occurs when all the essential functions cease. The key factors that determine death are irreparable damage to mitochondria, so energy can no longer be produced, and membrane breakdown. And as always with cells, knowledge of how they work can be surprising.

As well as evolving complex systems for multiplication and division, as well as means to correct errors, almost all of our cells have also evolved a mechanism for death by suicide. In our daily lives, bone marrow and intestine cells are dying all the time, in their billions. This death by suicide is quite different from death due to injury, which results in cells spilling out their contents and possibly causing inflammation. Death by suicide – known as apoptosis – is a death programme all our cells have, except for our red blood cells which do not have a nucleus. When this programme is activated the cell shrinks and the structures inside break down as a result of the action of special suicide enzymes. This breakdown gives a signal to the special white blood cells that clean up our tissues, and which now come and engulf them so their contents are never spilled out, and there is no damage to neighbouring cells – a noble action.

To think what might happen if they all did it at one time is terrifying. Fortunately the suicide programme only gets activated when cells are no longer needed, or if they present a danger to the organism by, for example, becoming cancerous, and death becomes a sensible and clear choice. Cells also commit suicide if they do not receive factors which ensure that they will continue to live, particularly during embryonic development. Many, many nerve cells die when they have not made the right connections during development, as we shall see. A classic example of programmed cell death during development is the death of the cells between our developing fingers, so that we do not have webbed hands like a duck's feet.

We now look in more detail at some of the many things that cells do, and particularly at the role of proteins, which are at the core of the mechanisms that drive these activities.

4

How We Function

how proteins determine the work of cells

Like factories with numerous assembly lines, cells have to do many things: grow and divide, move, and maintain their highly ordered structure. Other activities range from muscle contraction to conducting nerve impulses. All this is done by proteins.

Many of our different cells are organised into sheets – such as our skin, lungs and gut – and have different properties according to the function of the sheet. Just consider our skin, which protects us from bugs and dirt and keeps liquids from moving in or out. The outer layer of our skin is a continuous sheet all over our body composed of dead cells that fall off all the time, and their main protein is keratin, which makes them tough. Proteins between the cells make sure the fit is tight and that they adhere strongly to each other. The loss of the surface cells is replaced by stem cells beneath the surface dividing to give rise to the cells that replace those lost in the outermost layer. A different set of sheets with some different proteins makes up our extensive circulatory system, some 60,000 miles of vessels: the arteries and veins and capillaries. Our arteries branch to give 40 billion capillaries, the tiny vessels that join arteries and veins.

Membranes allow exchange of gases, food and waste, so that no cell is more than a few cell diameters away from the capillary which serves it.

Yet all our cells are composed of the same sort of molecules and share the same machinery for basic functions like making proteins. They also share a similar chemical composition, which is approximately 70 per cent water. Ninety-five percent of the mass of cells is made of just four types of atoms – carbon, hydrogen, nitrogen and oxygen – joined in a great variety of combinations to make almost all of the molecules in the cell. Joining these atoms together in so many different ways to form proteins and nucleic acids is what gives the cell its complexity. Combining basic units in different ways seems to be a key principle in the life of cells, which can generate enormous variety with relatively few basic units. And the key to cell function lies in the proteins.

Inside each cell is a society of molecules that can carry out all these activities, and the machines that do almost all the work are proteins, the most complex and varied of all molecules. The 200 or so different types of cells in our body – skin cells, nerve cells, liver cells, fat cells and many others – all have their function determined by the proteins they contain, and there are differences in the proteins made by different cell types. Later, we will look at how proteins are made.

The number of amino acids along a protein string is usually between 50 and 2,000, but there are proteins with just 30 amino acids and others with 10,000. Each of the thousands of different proteins has a unique sequence of amino acids, and this determines its function and behaviour. The functioning of proteins is largely determined by their interaction with other molecules, whose behaviour or structure they alter. Most interactions show great specificity: a particular protein can bind to only a few, or just one, of the thousands of other different molecules in the cell. This ability to bind with selectivity is due to the folding of the protein into an intricate three-dimensional shape, which results in sites being formed to which other molecules can bind. For example, the folding can produce a cavity on the protein's surface into which an interacting molecule can fit.

Proteins provide the cell with its structure and are the building blocks from which cells are built. They are a major component of membranes and also form the basis for movement by the cell. Proteins can also transport molecules across the cell membrane, in and out of the cell; they can act as signals between cells; they also act as receptors, and they control which genes are turned on or off. Proteins can co-operate and are capable of self-assembly by binding to each other to form filaments, microtubules, rings, sheets or even spheres. A major function of some proteins is their action as enzymes. Enzymes are proteins that bind to molecules and alter their structure by breaking them into smaller units or adding other molecules to them. They are thus fundamental in the formation of molecules and the breakdown of food into smaller units. The character and behaviour of a cell is thus determined by its proteins, and almost any activity of the cell is controlled by and carried out by different proteins.

The chain of amino acids of a protein is very flexible and allows the chain to fold in an enormous number of different ways – it makes a human contortionist look simple. The three-dimensional structure that it forms determines the protein's function. The way a protein actually folds is determined by the sequence of its amino acids, which can interact with each other both positively and negatively. Imagine a rope with hooks and rings attached to it along its length; by folding the rope randomly some of the hooks will attach to rings and change the shape of the rope – so it is with proteins. They quite often fold incorrectly and the cell then wisely destroys them, which uses a significant amount of the cell's energy reserves. To help proteins fold properly there are – in this well-ordered society – special proteins called chaperones.

The behaviour of the protein is very sensitive to the precise order of its amino acids. If just one amino acid in the chain is substituted by another the folding of the protein is significantly altered, and can result in serious alteration in its normal function and so lead to abnormal cell function. This is the basis of many illnesses, as we shall see. A clear example is sickle-cell anaemia, in which just one amino acid is altered in the protein haemoglobin. The mutation causes the protein chain to fold abnormally, so that it forms aggregates in the red blood cells. These deform the cells and give them a sickle shape, which makes their smooth flow through the capillaries difficult.

But how can the order of amino acids in a protein be determined? A well deserved Nobel Prize was given to Fred Sanger, a Cambridge biochemist, for determining the amino acid sequence of insulin. He developed a technique for chopping off the amino acid at the end of the chain and identifying it; he could then sequentially chop off amino acids one at a time all the way along the chain and thus determine their order. However, even knowing the order of amino acids in a protein chain is often not by itself enough to tell us what the protein's folded shape will be. A most successful approach has been to make crystals of the protein, and then shining X-rays through the crystal and seeing how they are deflected; it is then possible to determine the shape of the protein by analysing the pattern of the reflections. In practice, similarity with known structures is a far more accessible approach to prediction of protein structure, and is becoming more powerful the more structures we know.

How do proteins know where in the cell to go so as to perform their specific job? There are millions of protein molecules in most of our cells and some thousands of different types, and their future location is coded into their structure. Free-floating proteins probably meet thousands of other molecules, including other proteins, every second as they all dance about in the cytoplasm, and this can result in them finding their correct location when they attach to a specific site. But some proteins have to be specially delivered to those parts of the cell that require them, and it is also necessary to keep other proteins out of these regions so that they do not interfere. Membranes play a key role in keeping proteins in their place, as proteins cannot pass through membranes without special transport systems. The main membrane-bounded structures in cells that help proteins find their way around are the endoplasmic reticula, which is where proteins that are going to work in some parts of the cell move to first, and are told there where to go. This is done by modifying the protein, for example adding to it a sugar molecule which alters what it can attach to.

Enzymes are among the most amazing and versatile

proteins: they can convert molecules of one type into another, break down a molecule into its basic components, or join two molecules together. They are catalysts that can speed up a reaction a million times but remain unchanged in the process. They work at a remarkable speed, and they can act on about 1,000 molecules per second. They find the molecules they interact with by chance, but that is also amazingly often, as a particular enzyme will bump into the molecule it acts on some 100,000 times per second. The chemistry of the activities in the cell is, to put it mildly, very fast indeed.

Enzymes break down sugars which as we have seen, provide us with energy in the form of ATP. Enzymes also provide the means for the synthesis of many molecules, and their actions underlie the basic chemistry of the cell. One can think of an enzyme as the lock into which a key – the molecule that is to be altered – fits. It is a tight fit, and when they are brought together the enzyme changes the shape of the key and can break it down into smaller units, or can add another molecule to it. Enzymes often work in teams, the product of the action of one enzyme becoming the target of another enzyme, and this can lead to complex networks for making energy or synthesising new molecules.

Enzymes in our gut are essential for digesting the food we eat – that is, breaking it down into smaller units that can enter the bloodstream and be carried by it to nourish our cells. Another nice simple example of enzyme function is provided by lysozyme, which is present in our saliva and tears and can help prevent bacterial infection. This enzyme can sever the sugar chains in the outer wall of a bacterial cell, causing the bacterium to burst open and die, as the wall is necessary to withstand the pressure due to water trying to enter the cell. The way lysozyme works is by adding a molecule of water between two sugar groups in the sugar chain and causing the chain to break. The sugar chain is bombarded by water molecules all the time, but there is an energy barrier preventing them breaking the chain. The enzyme distorts the sugar chain so that the energy barrier to the chemical reaction is lowered and water can get in.

The activity or behaviour of proteins like enzymes is not autonomous, but is controlled by many other molecules so that their functions contribute sensibly to both the behaviour of the cell and its interactions with other cells. For example, a special feature of some proteins is that when they bind to one molecule, their shape is altered so that their interaction with another molecule is inhibited or improved. A common principle used to control enzyme activity is feedback inhibition. An enzyme produces a substance X that produces yet another substance Y, and Y can then feed back to interact with the enzyme to slow down its behaviour so that not too much of X and Y is made. This can be a very sensitive control system.

Another control mechanism widely used is the attachment of a simple phosphate group to one of the amino acids of a protein by an enzyme. This changes the shape of the protein and so how the protein interacts with other molecules. Removal of the phosphate group requires the action of an enzyme different from the one that added the phosphate group. The addition of a phosphate group to a protein is an important and common control mechanism, and at some time in their lives one third of all our proteins have a phosphate group added or removed to change their behaviour.

The component responsible for cell shape and movement is the cell's internal skeleton. This is made up of protein filaments and tubules, which are the bones and muscles of the cell, together with various membranes. A complex and dynamic network of protein filaments and microtubules – or sticks and rods – interact with each other to maintain the shape of the cell, and to cause it to change shape and move. The filaments strengthen cells against forces that stress and deform them; they form rope-like structures and distribute unwanted forces. In some cell sheets, like those in our skin, they attach to cell junctions and so give the sheet extra strength – clever engineering. The filaments also help support the membrane surrounding the nucleus assembled from small protein sub-units and provide the tracks on which small particles can move. Microtubules are relatively stiff hollow tubules that have the ability to rapidly disassemble and then assemble elsewhere depending on the local signals. Their relative instability makes them capable of rapidly remodelling, as we saw when we considered cell division, where they are involved in the separation of the chromosomes.

Cells, like us, have a well-defined internal skeleton and they are also highly mobile: if one examines a typical cell from the human body in a culture dish, one is struck by the constant movements inside the cell, as well as the cell's ability to move over the surface of the dish. Proteins are the material of which the cell skeleton is made and are also the machines that drive cell movement. An example is muscle contraction, which is fundamental to our lives. When our muscles contract they get shorter, and you might expect that the contractile forces would result from shortening of some molecules, but this is not the case. Cells have found a cleverer way.

This is most clear in our skeletal muscles, the muscles attached to the bones of the skeleton which are composed of chains of tiny contractile units, the muscle engine. The long fibres of our skeletal muscles are each a huge single cell, formed from the fusion of smaller cells during development. The contractile unit itself is very short, but there are many of them in a chain running down the length of the muscle. These units are made up of protein and the protein filaments myosin and actin. In each unit myosin and actin filaments are arranged in parallel bundles in such a way that the actin filaments lie between the myosin filaments. Contraction results from them sliding past each other, so making the contractile unit shorter – it is rather like sliding the fingers of one hand between the fingers of the other. Special regions on the myosin filaments pull on the actin, making it slide past, and this requires ATP. Because of this ability to power movement, myosin is called a motor protein. There is in fact no shortening of the actin and myosin filaments as a muscle contracts – just sliding.

Actin is a versatile protein, and can easily assemble into filaments and then break down again into its basic sub-units. It can form stiff structures as well as being able to form contractile bundles with myosin, which are part of the contractile ring that divides an animal cell into two at cell division. Actin plays a key role in enabling cells to crawl over a surface – typically shown in the movement of white blood cells when they come into tissues to eliminate invading bacteria. At the front end of a migrating cell there is a dense collection of actin filaments, and their growth can press against the cell membrane causing a small local extension, which sticks at its tip to the surface just ahead of the cell. There is some evidence that these extensions then contract, so pulling the cell forward, and then repeat the process – like climbing a ladder with only one's hands. But there is also contraction at the rear end of the cell based on myosin.

The continual motion seen inside the cell is mainly due to the movement of particles such as mitochondria and small membrane-enclosed vesicles, which move in small jerky steps. It is based on the same mechanism used for muscle contraction. Both filaments and microtubules are the tramlines along which these movements take place. The engine for these movements is another set of motor proteins, which bind with one end to the membrane of a vesicle or mitochondrion and then 'walk' with it along a filament or microtubule. The walking involves two 'feet', which successively bind and unbind to the tubule, like the action of myosin pulling actin past it in muscle contraction.

Motor proteins are also responsible for the movement of cilia. Many of our cells have a single hair-like structure extending from the surface – a cilium. The cilium bends and sweeps around like a flexible rod to move fluid over the cell. Cilia are involved, for example, in keeping the lining of our lungs clean: billions of cilia are continually sweeping up dust from our lungs into our mouth. The human sperm swimming towards the egg after ejaculation is propelled by a structure very similar to a cilium but larger. It waves and so drives the sperm forwards rather like a swimmer with flippers on. In all these waving rods the main structures are nine pairs of microtubules arranged in a ring which, though quite stiff, can be bent, a movement due to the microtubules sliding past each other. This sliding is brought about by motor proteins. It has only recently been recognised that many more of our cells have a cilium than previously recognised, and that they may be involved in cell signalling.

Proteins also perform crucial functions on the outer membranes of cells. Sugar molecules attach to the outer surface of the cell membrane, providing an extra-cellular coat that helps prevent both mechanical and chemical damage, but also on the outer surface are proteins that can bind to proteins on other cells, resulting in the cells adhering to each other. There are also protein receptors for proteins that arrive as signals on the outside of the cell membrane. When the signal arrives from other cells, the receptor can transmit the signal into the cell and so tell the cell, particularly the genes in the nucleus, what is happening in other cells. The passing on of such signals often involves an apparently eccentric sequence of protein interactions.

Cells need to know both what their neighbours have to say, and must also listen to long-range chemical signals like hormones that might be sent out from distant sources. For example, insulin signals to cells to allow sugars to enter. Because of their central role in interacting with the cell's outer world, the number of proteins required for membrane function in all our cells may be as many as 10,000, a significant fraction of all the different proteins in our body.

The basic structure of the cell membrane is made up from fatty molecules – or lipids – and protein molecules, and is based on the fact that fatty molecules dislike water. The lipid molecules do not mix with water, and always try to keep separate from each other, and so they are arranged as a very thin double layer; this gives the membrane its fluidity and makes it largely impermeable to water soluble molecules as glucose. The lipid molecules in the cell membrane are like tiny animals that hate water: their head end does all it can to keep out of the wetness, while the tail end can just cope with contact with water. Proteins embedded in this bimolecular lipid layer make up about half of the membrane's composition. The protein molecules in the membrane lie across or in the lipid layer, and are responsible for movement or the transport of molecules across the membrane.

Our cell membranes are thus quite fluid and flexible when the cell changes shape, and will not burst open even if pierced. New membrane is added by vesicles, tiny membrane-bounded hollow spheres, or sacs, inside the cell and fused with it. The ease with which lipid molecules form double layers may have been important in the evolution of the cell membrane and the cell itself.

In spite of lipids disliking water, water can slowly enter or exit across the membrane. The lipid layer allows only water and very small molecules that do not have an electric charge to move across it. Water-soluble molecules like oxygen and carbon dioxide can also diffuse rapidly across the membrane. By contrast, any molecule that has an electric charge, such as sodium and potassium ions, has great difficulty getting across. An ion is an atom or molecule that has lost or gained one or more electrons, making it positively or negatively charged. When ions need to get in and out of the cell there is a protein transport system for them in the membrane. For larger molecules, proteins in the membrane also provide the necessary permission. The membrane has two transport systems, with some proteins providing a channel and other proteins acting as a carrier or providing active transport.

The sodium ion has a concentration outside most of our body cells that is nearly twenty times the concentration inside the cell, whereas the concentration of the potassium ion is just the opposite. These differences in concentration are due to the sodium pump, a protein which pumps sodium out and potassium in. This sodium pump is crucial for stopping the cell bursting, as there is osmotic pressure driving water into the cell to equalise the concentration of the many molecules within it with the lower concentration outside the cell. It is only by pumping out sodium and reducing its internal concentration that the cell does not burst from the pressure of the water entering. Block the action of the pump and the cell will burst and die. Nearly one third of the cell's energy – that is, one third of all your energy – is needed to drive this pump. It may seem extraordinary that so much of your life is devoted to feeding the sodium pump, but the pump is essential and keeps us alive.

The sugar glucose, which is essential for the cells to produce energy, needs a special mechanism to enter, and this requires the protein insulin. In the presence of insulin glucose is ferried through the cell's membrane by a family of molecules known as glucose transporters. The transporters are contained in tiny sacs, or vesicles, inside the cell; they are transferred from the vesicles to the cell membrane when insulin binds to the membrane. The vesicles travel along a network of microtubules and then attach and fuse with the cell membrane; the transporter, contained in the vesicles' membrane, then enters the cell membrane, where it ferries glucose into the cell. Failure of this transport mechanism results in diabetes, as discussed later. There are also carrier proteins for other essential molecules, such as the amino acids used to build proteins, and the nucleotides used to build nucleic acids like DNA.

To see what determines the order of the amino acids in proteins we must look next at genes. Contained in the nucleus of our cells, genes are the blueprints for proteins, and as such they determine protein structure and our heredity.

5

How Genes Work

how DNA encodes proteins

Each of our cells has some 30,000 genes, while some bacteria can manage with fewer than 500. They provide the code for making proteins by specifying the sequence of their amino acids. You have the same genes in the cells around your anus and in your brain, and for all different body structures, but different genes are turned on in each type of cell, providing the code for the different proteins made in, for example, our skin, nerve and muscle cells.

In our 46 chromosomes there are an astounding 3,000 million nucleotide pairs such as G–C and A–T, as described earlier. If the DNA from every cell in a single human body was stretched out to its full length, and then all these strands were joined together, how far would it stretch? The answer is 200 times from the earth to the sun and back. Another way of thinking about it is to imagine folding 40 kilometres of fine thread into a tennis ball. Within the cell, much of the DNA is packaged into a dense form by proteins.

The cells package their long DNA sequences in a variety of ways in the chromosomes, and this compaction can increase 10,000 times when the chromosomes prepare for cell division. Special proteins are involved in the packaging of chromosomes, and the DNA is wrapped around particles of this protein, with DNA providing the link between the particles. There are regions that vary in how densely they are packed, and there are other proteins that can undo the packing and make the DNA more accessible to being decoded.

The As, Ts, Cs and Gs in the long DNA molecules are arranged in a specific order to provide the code for proteins, which are made of twenty different types of amino acid sub-units. The way in which DNA acts as a blueprint is that for each protein there is a gene, a stretch of DNA which codes for the sequence of amino acids in that protein. Gene language is translated into protein language: a particular sequence of nucleotides determines the order of amino acids in proteins, telling the proteins their basic nature. The system is rather like the Morse code, where dots and dashes code for each of the letters in the alphabet. The sequence of nucleotides, read three at a time, corresponds to the sequence of the amino acids along the protein. Each set of three nucleotides codes for one amino acid; for example, AUG codes for the amino acid methionine. There are 64 combinations of the four nucleotides taken three at a time, but there are only twenty amino acids, so there are triplets that are never used to code for proteins but are used instead as punctuation marks – to indicate the end of a protein-coding sequence. No triplet is without some function, while there are several triplets that code for the same amino acid. Each of our proteins thus has a corresponding gene that codes its sequence. The largest gene has some two million nucleotides along each strand, and the smallest around a thousand.

Our DNA is a precious information store that remains inside the cell nucleus. But proteins are synthesised not in the nucleus but outside it, in the cytoplasm. How can this happen? First the DNA code for a protein is transferred to another nucleic acid, messenger RNA. This messenger RNA is, like DNA, a string of four nucleotides, but unlike DNA it is single-stranded (the DNA in our chromosomes is always in the form of a double-stranded helix). Another difference is that instead of the base T, RNA has uracil, which happily binds to A. This means that the sequence of an RNA can be complementary to that of DNA. The discovery of messenger RNA came from the realisation that there must be a means of carrying the information for making a protein from the DNA to the site of protein synthesis in the cytoplasm. This occurred to Sydney Brenner, a South African scientist and my personal hero, and Francis Crick during a discussion with others on Good Friday 1960 in Cambridge. Brenner then went with to the USA to do the key experiments. When the experiments were going badly wrong they went for a rest on the beach. During the discussion Brenner sprang up shouting, 'It's the magnesium!' His wide knowledge led him to realise that they had not added enough magnesium to the experimental solution. They went back to the lab and discovered messenger RNA.

The 'switching on' of gene activity refers to transferring the code to messenger RNA by a protein machine that copies the code on the DNA, the sequence of the nucleotide bases. This process, called transcription, begins with the opening and unwinding of a small portion of the double helix of DNA at the end of the gene, the promoter region. The sequence of coding bases of the gene is then copied from one of the DNA strands, one by one, on to the growing messenger RNA molecule as the machine moves along the DNA strand. In this way a complementary coding sequence is transferred from the gene to messenger RNA. The process ends at another recognisable sequence, the terminator. One might think that the messenger RNA at this stage is now ready for translation into the amino acid sequence of a protein, but life in the cell is, as usual, much more complicated.

In cells such as ours, most genes contain many more nucleotides than those required to code for the protein. These are present in stretches of non-coding DNA called introns, which interrupt the coding sequence. They are copied onto the messenger RNA, but have to be removed from it before it can act as a proper message for making a protein. The regions that actually code for the protein are known as exons. Only when the introns have been removed by a clever mechanism called RNA splicing, which leaves the exons intact and joined together, is the messenger RNA allowed to leave the nucleus and pass into the cytoplasm. There is one further complication. When the RNA is being spliced to remove introns and join up the exons, in many cases the exons can be joined up in more than one way and so code for different proteins: this alternative splicing is strictly controlled so that the right protein is made in the right time and place in the body. Alternative splicing is the reason that there can be about twice as many different proteins in our body as there are genes.

Whether a gene is switched on and transcribed into RNA depends on the binding of special proteins – transcription factors – to particular control regions in the

DNA. These control regions do not encode proteins but provide recognition sites for the transcription factors and are required for the protein machine that transfers the code to messenger RNA to function. A promoter region, which is where transcription begins, is present just before the coding region, while other control sites may be near the gene or far away. Only if the correct control regions are occupied by the right transcription factors can a gene be transcribed by the protein machine which copies the nucleotide sequence.

The control of gene activity can involve many transcription factors, and they bind to the DNA control regions for a particular gene. The control regions are outside the coding region and there may be several of them, some a long way away from the coding region. It is the transcription factors that hunt and find control regions.

We talk about a gene that is being translated into messenger RNA as being turned on, and those for which there is no such decoding as the gene being off. When the correct transcription factors bind to the control region, the gene begins to be translated into messenger RNA. There is a control region near the start of the gene, but the control region may also involve DNA sequences a long way away and only if transcription factors bind to such regions of the DNA will the protein transcription machine begin its work. Regions of the DNA thousands of nucleotides away can be involved. Transcription factors may bind to this distant site in the DNA, and then the DNA folds to bring the factor to join the others in the control region close to the gene, and so the gene is turned on. The gene will remain active so long as the control regions are activated.

There can be more than one set of control regions for a particular gene, for its activity might be required in quite different situations. The importance of these control regions cannot be overemphasised; we will come back to them later when we look at the development of the embryo. The protein produced by one gene can activate several other genes or even inactivate them, and so a system of gene interactions is set up which determines cell behaviour and how it changes with time.

The spliced messenger RNA leaves the nucleus, enters the cytoplasm and goes to a ribosome. Ribosomes are small round protein synthesising machines where the sequence of nucleotides of the messenger RNA is translated into the sequence of amino acids of the protein in the order determined by the bases along the messenger RNA. The actual translation of the messenger RNA into a protein by the ribosome is based on small RNA molecules known as transfer RNAs, which can recognise a coding triplet and at the same time bind to the amino acid for which the triplet codes. For example, the amino acid lysine is coded for by both AAA and AAG, while tyrosine is coded for by UAC and UAU, and these are recognised by transfer RNAs. The ribosome then acts as a protein-making machine. One of the most complex structures in the cell, it is composed of both RNA and protein. The ribosome machine moves down the messenger RNA and joins up the amino acids that are presented to it by the transfer RNAs. Ribosomes work fast, as is so common in cell processes, and in one second join up two amino acids along the assembling protein. Proteins are formed in times ranging from 20 seconds to several minutes.

We can now understand the nature of a mutation and how it can affect a cell's behaviour. A mutation in the DNA can alter the sequence of bases in the coding region of a gene, and so alter the normal sequence of amino acids. This can alter how the protein folds and functions and can result in a faulty protein being made, which can have serious positive or negative consequences as we shall consider later on. Mutations that alter protein function in egg or sperm cells are the basis of evolution, as the mutation will be passed on to the next generation. Changes in the DNA in the control regions can also affect cell behaviour as they determine when and in what cell a gene is made active and can be translated into RNA.

The DNA of humans has been sequenced, all the 3,000 million base pairs. The average size of a gene is about 27,000 base pairs, but – and this is a big but – only about 1,300 nucleotide pairs in these genes actually code for a protein of about 430 amino acids. The difference between two people who are not related is less than 1 per cent, but that involves some three million nucleotide differences.

It is hard to comprehend all this information in most of our cells. I was struck by an exhibition at the Wellcome Trust where the whole human sequence was present for inspection in a large bookcase filled with thick volumes. Opening any of these volumes on any page gives one a shock, for all there is, in small print, is the sequence of nucleotides ATGCTGACCGATTAGTCA . . . on and on and on and on, for some five hundred pages in each volume. Just looking at it fills one with awe for the capacity of cells to store and use this information. The sequence in the cell runs to two metres in length – so how does the cell find the right gene? How does a cell know in which of those books and on what page the gene they need to turn on is to be found? What determines whether a gene is active and thus whether messenger RNA for a particular protein is made? The answer to that question is a fundamental feature of the control of cell behaviour, as it determines which proteins are present in the cell and thus also its behaviour.

We can now see what genes do – or really do not do, for in the whole process of protein synthesis they are passive. They only encode proteins and have control regions that turn them on. And when it comes to gene replication, it is again proteins that do the work. It must always be remembered that DNA is not a blueprint for the organism. as will become clearer when we look at how the embryo develops. Nevertheless it is striking how many different cell states the genes can specify by the proteins they code for. There are in our cells some 30,000 different genes and if we consider that a few hundred different proteins could specify a particular cell state then, with different combinations of those genes being turned on, there are billions of possible different cells, each with a unique combination of proteins.

The structure of genes and the role of DNA are now being recognised as becoming even more complex, frighteningly so. Only a small percentage of the DNA in our chromosomes actually codes for proteins, less than 2 per cent. There are many repeated regions whose function is not known. Much of the DNA has thus been thought of as just accumulated junk DNA. Of course some of it is a control region, determining when and where a gene is turned on. But DNA is still exceedingly complex.

Stretches containing repeated sequences constitute as much as 10 per cent of the chromosomes, and some of these sequences move around from one place to another. This jumping involves a stretch of DNA being copied and then inserted elsewhere. Many mutations are due to these jumping sequences landing in a coding or control region and so changing its character. The DNA for a gene can also be much more complex than previously thought. There is new evidence that some genes can have their DNA sprawling, with their protein coding regions overlapping with other genes a long way away. Why, we do not know.

For a long time RNA has been regarded as a humble carrier of messages from the genes to the ribosomes, where the proteins are synthesised. But quite recently it became clear that RNA is doing all sorts of other things. It now has been discovered in a major collaborative worldwide effort that there is a great deal of transcription of the DNA into RNA of unknown function. Many different RNAs are transcribed from the DNA that are not related to protein manufacture. Indeed there are claims that there are more such transcripts than there are genes. These micro-RNAs, as they are called, can regulate the function of messenger RNAs and whether they get translated into protein, and some individual micro-RNAs control the concentrations of hundreds of different proteins in the cell. Micro-RNAs are just some twenty nucleotides long, and one of their main functions is to bind to messenger RNA and destroy it so that protein is not made. They clearly provide the cell with yet another control system for the synthesis of particular proteins.

All proteins are synthesised on ribosomes in the cytoplasm of the cell. Some ribosomes are attached to structures like the membranes of the endoplasmic reticulum. Proteins manufactured by these ribosomes are fed into the endoplasmic reticulum, and enter a postal system that sends them to the places where they must work. The way the proteins are transported again involves the membranes. It is a complex but very efficient postal system. There is a further messenger service for messenger RNA itself, which is transported to the site where its protein is required and only when it arrives there is the protein synthesised. When a cell is migrating, actin is required at its front end in quite large quantities in order to form the filaments that push the front forward. The messenger RNA is transported to the leading edge and the protein is made there. Just how it is transported and prevented from being synthesised until it arrives remains to be properly understood.

Cells can also tell the time. We are familiar with the fact that our bodily functions are all too dependent on a daily rhythm, which is upset when we travel by air east or west. It is a rhythm that affects not only our sleep–wake cycle but also hormone secretion and both liver and kidney function. Light intensities set this circadian 24-hour clock. When light reaches the retina in our eyes, signals are sent to a special region of our brain that is the major timekeeper in the body. But it is not just in your brain that there is a clock: it now seems that every cell in the body may have a 24-hour clock ticking away. In tissues as different as liver and lung there are genes that oscillate by being turned on and off with a daily cycle. Cells that have been in culture for twenty years can be made to oscillate again. It is not inconceivable that the clock-like activity in our cells may be more relevant to our health than previously thought. One need only recall the sleep problems of depressed patients and how awful jet-lag can be. Perhaps in the future there will be clever ways to reset all the clocks in all the cells.

Everyone has a unique set of genes, unless they have an identical twin. Analysis of DNA can thus be used as a way of identifying a person, a powerful tool in criminal cases and the identification of accident victims. But what could one tell about someone if their DNA was sequenced? Would it tell you everything about that person's body? It could certainly tell you if they suffered or might suffer in the future from well-known genetic diseases, which usually involve mutations in more than just one gene, but little more.

Damage to DNA by the alteration of just one nucleotide can have serious effects because it changes what proteins are made. Change in just one nucleotide in the gene for haemoglobin will result in sickle-cell anaemia: the protein will fold incorrectly and will deform the cell into an abnormal sickle shape. Replication is the main time at which mutations, or changes in the nucleotide sequence of the DNA in the chromosomes, occur, but it is not only at this time that DNA may be damaged and lead to a mutation. DNA, like all the other molecules in the cell, is continually being bombarded by other molecules hoping to meet their appropriate target. Unfortunately, some of these collisions can chemically alter the DNA in such a way that one or more nucleotides are replaced by different ones, changing the coding. Such changes often mean that the encoded protein will not be normal and will not work properly. Ultraviolet light, radiation and some chemicals can also cause changes resulting in a mutation, and DNA is constantly bombarded by collisions with other molecules. But cells and evolution are all too aware of this, and there are repair mechanisms available which can work if only one strand of the DNA is damaged. The strands are separated by special proteins in the damaged region and the correct nucleotide replaced.

The DNA in our chromosomes is not as stable as one might have expected. Repeated sequences can jump around the chromosome. And there is a further aspect of moving genes – viruses. Viruses are essentially genes enclosed in a protective protein coat. Viruses are not alive and can only reproduce inside a cell, where they use the cell's machinery for replication. They are basically parasites. Their replication is often lethal for the cell, which breaks down and releases the virus to infect more cells. Their small genome codes for their protein coat and for enzymes to hijack a cell's replication machinery. The simplest virus has a coat made of a single type of protein enclosing just three genes and is the cause of German measles. More complex viruses have several hundred genes and an elaborate shell, and we will turn again later to how viruses cause illness. The 'genes' in some viruses are made of RNA, not DNA.

Genes control development of every bit of our bodies by determining which proteins are present in our cells, but I have little sympathy with those who talk about genes for this or that, such as good looks or intolerance. Whether or not such characteristics have a genetic basis is a tough problem. The language is misleading. There is no gene, for example, for the eye; many hundreds, if not thousands, are involved in its development, and a fault in just one can lead to major abnormalities.

The language in which many of the effects of genes is described leads to confusion. No sensible person would say that the brakes of a car are for causing accidents. Yet, using a convenient shorthand, there are numerous references to, for example, the gene for homosexuality or the gene for criminality. When the brakes of the car, which are there for safe driving, fail, then there is an accident. Similarly, if criminality has some genetic basis, which is by no means certain, then it is not because there is 'a gene' for criminality but because alterations in many genes have resulted in this particular undesirable effect. Such alterations could, for example, have affected how the brain developed, or could cause changes in the way the nerve cells of the adult brain function.

It is important to realise that at present, knowing all the genes of an individual will not tell us how the cells function, or how the embryo develops. Consider this set of words: 'alters impediment marriage me of not not or to true is minds with love love which let it alteration the finds bends remover remove admit when the to'. Taken from the beginning of a famous sonnet by Shakespeare, these words as printed suggest the subject matter of the poem – it seems to be about love and marriage – but give no idea whatsoever of the ideas expressed. This is the position we are in at present in relation to the human genome. We know the whole DNA sequence of the human genome – that is, the DNA sequence of all human genes – and it is a tremendous achievement to have sequenced all that DNA, because by it we are able to identify each of the 30,000 or so genes that control our development and our lives. But each of those genes is like a word in the poem, and by itself tells us little about how the protein it codes for will act in the body or how it will interact with other proteins. Knowing the human genome sequence is only the beginning, and there is an enormous amount that it cannot tell us. We need to know how it controls the development of the embryo; then we would know how the sonnet reads:

> Let me not to the marriage of true minds
> Admit impediment. Love is not love
> Which alters when it alteration finds
> Or bends with the remover to remove:

Just six years ago, two versions of the human genome sequence were published. Both these versions are composite sequences derived from the genomes of many anonymous donors. But now one of the principals behind this work, the American scientist Craig Venter, has sequenced his own genome. No one expects to understand what makes Venter tick by looking at his genome, but there are some potentially useful things one can find out from a genome sequence, such as looking for known mutations that increase the risk of various diseases.

Will it not be possible to work out from the genome just how the whole will function? Perhaps in the future, but not at present, any more than we can reconstruct the sonnet from its words. The key to understanding living systems is proteins, and genes merely provide the information for making proteins. Knowing the human genome will tell us what proteins cells can make, because the coding regions can be identified, but it will not easily tell us when or where they will be made in the developing embryo. This depends on the control regions of genes, which cannot be identified in the genome without much more research. In addition, there will be proteins whose function is still unknown.

It will also be extremely difficult to work out how many of the proteins interact with one another. Tens of thousands of proteins are involved in constant communication with each other within the cell, and the genome on its own may never give the sequence of interactions. We may know that the words come from Shakespeare's 'Let me not to the marriage of true minds admit impediment', but not by analysis. The future will be the age of research into determining what the sequence of nucleotides really tells us about cell behaviour and the complex interactions in the cells, from DNA to proteins. And what the proteins do.

6

How Our Cells Are Replaced

how stem cells self-replicate

Stem cells are self-renewing – they can divide again and again, giving rise many other cell types. Every day they are dividing in our bodies, giving rise to blood cells, skin cells, cells lining our gut, the cartilage precursors of bone growth, and even some nerve cells and nerve-supporting cells. All these cells are in situations that require replacement, as in the skin and gut where cells are lost continuously. After dividing, one of the daughter cells remains a stem cell while the other can develop into a specialised cell such as a skin cell. They can also divide symmetrically, so that both daughter cells remain as stem cells.

Because of their capacity for an enormous number of cell divisions, as well as giving rise to different cell types, stem cells provide an exciting and promising way of treating a variety of health problems. They could be the core of regenerative medicine. The stem cells with the greatest ability to develop into all types of cells are those that can be taken from the early human embryo – human embryonic stem cells. This has raised many ethical issues; many people who, often on religious grounds, regard the early human embryo and even the fertilised egg as already a human being object to it being manipulated or cells being removed from it. A further ethical issue arises with cloning: it is possible, in principle, to create an embryo with the genes of a particular individual by transplanting a nucleus from a body cell into an egg whose nucleus has been removed and thus obtain stem cells with the genes of the donor. These recent advances in cell biology have given rise to great controversy at the same time as opening up new opportunities – it will very likely be possible, for example, to replace nerve cells in patients with Parkinson's disease or to replace cells in a damaged heart using stem cells.

All this current excitement about stem cells requires us to understand how the fertilised egg develops. Unless the process of development is understood, it is difficult to judge whether stem cells and cloning really do raise ethical problems. I will also try to set out the various misunderstandings that have led to these issues becoming so contentious.

The first stem cells discovered were those for blood. One of the puzzles in the early 1960s was why radiation was helpful in treating some cancers, and how the new technique of bone-marrow transplantation was able to restore blood systems whose cells had been destroyed by radiation. Two Canadian scientists, James Till and Ernest McCulloch, began by injecting varying amounts of bone marrow cells into mice after the animals were given a lethal dose of radiation. 'We found that the more marrow cells you gave, the higher percentage of animals survived,' Till recalled. To figure out how the transplants saved the animals, they began a series of careful observations of the blood-forming tissues of mice after injections of bone-marrow cells. In the course of these experiments,

McCulloch observed odd bumps in the spleens of some of the animals. 'Both of us had experience in cell culture – we immediately thought colony,' Till recalls. With the help of graduate students they showed that the strange bumps were indeed colonies of a sort – they were clumps of cells somehow derived from the transplanted bone marrow. They developed a definition of stem cells that holds true today: they are self-renewing, able to give rise to differentiated descendants, and capable of extensive proliferation. They and their colleagues then showed that a single stem cell from the marrow could produce any type of adult blood cell.

Transplantation of bone marrow into patients with illnesses like leukaemia, after the stem cells which give rise to the cancer are destroyed by radiation, have been quite successful, and show the value of the stem cells that these grafts contain. Rather few stem cells are required to re-establish the blood system, but a problem is immune rejection of the injected cells, even when the cells come from a close relative.

All our blood cells come from blood stem cells in our bone marrow together with those formed in our livers when we were embryos. These stem cells divide, and one of the daughter cells can go on to give rise to one of the several types of blood cells that we have, in particular our red blood cells. The stem cells also give rise to the cells of the immune system, and a variety of other cell types in our blood – such as the macrophages that engulf any fragments lying around, including invaders like bacteria and viruses, and also the platelets which are important for clotting of the blood when there is a wound. The blood stem cell also gives rise to cells that can break down bone when remodelling is involved during embryonic development or when there has been damage.

The main cell type that the blood stem cell gives rise to is the red blood cell, which contains the protein haemoglobin that carries oxygen from our lungs to our tissues. Red blood cells have a short life of only about 120 days, and it is an extraordinary fact about our society of cells that they make some two million red blood cells in our marrow every second. The red blood cell cannot divide, as during its development its nucleus and mitochondria and much else has been pushed out, and all that really remains is haemoglobin. Their extruded contents are digested by macrophages.

How does the stem cell know which kind of blood cell to give rise to and how do these different types develop? The stem cells already have active some of the main genes that are required in the types of blood cells that they will give rise to, and so differentiation of the cell types involves inactivation of some genes and activation of others. This is controlled by some 200 transcription factors that control gene activity. Extra-cellular proteins close to the stem cell control how much the cells proliferate and which cell type will develop from the stem cell. The region where stem cells reside and where their fate is controlled is known as their niche. There are also signals from a distance; for example, when there is a lack of oxygen production of red blood cells is stimulated by a hormone made in the kidneys.

The outer layers of our skin are replaced around a thousand times during a normal lifetime. This replacement of the cells is based on stem cells that give rise to skin cells and remain capable of repeating the process again and again. Skin stem cells occur in the deep layer of the skin and at the base of hair follicles. The skin stem cells give rise to keratinocytes, our outer skin cells, which migrate to the surface of the skin. There they die and become flattened but packed with the protein keratin, and form a protective layer. These dead cells fall off our body all the time and are the source of much household dust. The follicular stem cells can give rise to both the hair follicle and to the epidermis. Again, in our intestine the covering layer is renewed almost every week, stem cells providing the new cells. Another example is the cells in our nose which detect odours. There are also stem cells in certain parts of our brain that can give rise to nerves in adults. Our muscles have stem cells close to them, and if a muscle is damaged these stem cells can differentiate into new muscle cells.

The way one of the two daughter cells is specified to remain a stem cell is not yet fully understood. The stem-cell niche is of great importance. One theory is that signals from the surrounding cells in the niche can determine both the nature of the stem cell divisions and how the daughter cells will then develop; the niche cells can tell which of the daughter cells after division is to remain a stem cell. The mechanism may, however, be quite different and be based on the distribution at division of a set of special proteins to only one of the daughter cells when the cell divides in two. It is also possible, as already mentioned, for a stem cell to divide so that both daughter cells are stem cells, like identical twins, which is a means of increasing the number of stem cells at a particular site.

The stem cells responsible for tissue renewal in adults, like those for blood, gut tissue, skin and nerves, can generate only a limited number of types of cells. They can also operate only in very specific environments, like blood stem cells in the marrow; they need a special niche, and though blood stem cells are present in our blood, they cannot function there. But could these stem cells be persuaded to give rise to cell types different from their normal behaviour? Could blood stem cells give rise to nerve cells? There have been claims that such stem cells can give rise to cell types other than those which they normally produce. There was a report that nerve stem cells, when transplanted into mice whose bone marrow had been destroyed, gave rise to blood forming cells again in the marrow. But it has not been possible to replicate this finding. Other studies reported that after culturing bone-marrow cells they behaved as if they could differentiate into cell types not related to blood, but again it has not been possible to confirm this. Claims that blood stem cells could repair liver damage due to a faulty gene in the liver were in fact due to the injected cells fusing with the liver cells and so providing them with a healthy gene replacement. But embryonic stem cells do not have such problems.

Embryonic stem cells are very special compared to those involved in renewing cells in the adult. Their potential to develop into different types of cells is almost unlimited; in fact every one of our cells is derived from these cells in the early embryo. Embryonic stem cells obtained from the inner group of cells of the early embryo can give rise to all the different cells in our body, and they are thus termed pluripotent. If these inner cells are placed in a culture dish and provided with normal nutrients they will apparently proliferate indefinitely, dividing every twelve hours. Moreover, and this is crucial, they retain their pluripotency. By manipulating the culture conditions in which the embryonic stem cells are growing, it is possible for them to be made to differentiate into particular cell types like nerve, blood and muscle. Mouse embryonic stem cells can even give rise to eggs. Just how this pluripotency is maintained is only partly understood. To prevent the cells differentiating into different cell types, and to maintain their ability to proliferate, certain genes, which have been identified, and culture conditions are essential. In most experiments taking stem cells from the early embryo damages or kills the embryo. However, it may be possible to take just one cell from the early embryo that will give rise to stem cells in culture, and do no harm to the embryo.

In principle, these embryonic stem cells could be used for regenerative medicine by restoring tissue function when particular cell types are lacking or damaged. Perhaps the most important potential application of human stem cells is the generation of cells and tissues for use in cell-based therapies. Today, donated organs and tissues are often used to replace ailing or destroyed tissue, but the need for transplantable tissues and organs far outweighs the available supply. Stem cells, when directed in culture to differentiate into specific cell types, offer the possibility of a renewable source of replacement cells and tissues to treat diseases including Parkinson's disease and Alzheimer's disease, spinal cord injury, stroke, burns, heart disease, diabetes, osteoarthritis and rheumatoid arthritis.

Parkinson's disease is due to a gradual loss of a specific set of cells that use the chemical dopamine as signal to other nerves. It results in tremor and instability. Transplantation of embryonic cells producing dopamine has caused significant improvement in the symptoms. However it has not yet been possible to transplant nerve cells of the type lost in the patients from stem cells, and so restore full function. Transplantation of stem cells into patients with spinal cord injuries that lead to a loss of movement and sensation are encouraging, even if the detailed mechanisms are not yet fully understood. Nerve-cell replacement is an urgent problem, and there is much research in this area.

Experiments on mice have suggested that it will be possible to use embryonic stem cells to replace the muscles that have degenerated due to muscular dystrophy, as well as those in the heart that have died during a heart attack, and to cure Type1 diabetes by replacing the insulin-secreting cells. This would depend on the stem cells giving rise to cells that characterise the region to which they are transplanted, which would have to provide a suitable niche. If this works, as it does for transplants of cells from the marrow, they could be the replacement for transplantation of organs like the pancreas. Other recent studies have found that stem cells from the lungs of mice, when injected into the tail, all went to the lung, suggesting that this could be a means of treating certain lung malfunctions.

Encouraging as such experiments are, it is important to remember that studies on mice show that embryonic stem cells, if injected into the skin of a mouse, will give rise to a tumour made up of many different cell types. Stem cells must be injected into humans with appropriate care. There is also the problem that the immune system will reject stem cells taken from someone else, as they will be regarded as foreign.

One way to avoid immune rejection would be by modifying the genetic constitution of the embryonic stem cells so that they are not treated as foreign by the body. Nuclear transplantation can do this by putting the nucleus from one of the patient's cells, even from a connective-tissue cell, into an egg, which is then allowed to develop. If this method works then it would be possible to isolate embryonic stem cells from an early embryo which have exactly the same genetic constitution as the patient. These embryonic stem cells would not cause an immune reaction. However, although there has been recent success with both humans and other primates, it has proved difficult to get human eggs to develop after nuclear transplantation. And there remains the problem that the genes in the cells obtained by cloning from the patient may include those responsible for the original problem.

Another way of overcoming the problem of immune rejection may be achieved if pluripotent stem cells could be directly derived from patients' normal body cells. There has fortunately been a major advance. The genes that must be on for a cell to have the characteristics of an embryonic stem cell have been identified – there are only four key genes. It has been shown that cells similar to embryonic stem cells can be generated from mouse and human skin cells by using a virus to introduce these four genes that code for protein transcription factors. The transcription factors turn on the genes necessary for the cells to become like stem cells. This has led to the generation of high-quality cells from body cells that are comparable to embryonic stem cells in shape, proliferation, and gene expression. The high quality of these cells underscores the possibility of using this technology to generate patient-specific pluripotent stem cells. However there is some evidence that these cells may result in tumour formation because of the use of a retrovirus to put the transcription factors into the cell, and this requires further investigation. Much work is still needed to understand the molecular pathways of reprogramming normal cells to turn them into stem cells, and to eventually find molecules that could achieve reprogramming without transfer of potentially harmful genes such as those present in a virus. If successful, it would be equivalent to turning lead into gold. There are also encouraging new studies which show that it is possible to make embryonic stem cells from the testis.

Recent advances in cell biology have not only raised ethical issues and prompted objections from religious groups but have also got caught up with patenting and the law. In 1988 scientists at the University of Wisconsin isolated and cultured human embryonic stem cells, which they then patented together with the associated techniques. Human embryonic stem cells are hard to obtain because of the limited supply of human eggs, and many in the stem-cell community were most unhappy about this development. In April 2007 the patent office in the USA revoked its earlier decision, but unfortunately that decision has now been overturned and the patents have been confirmed.

A way of getting round the shortage of human eggs to make embryonic stem cells for experimental studies is to produce chimaeras with other animals. It is possible to make human-like embryonic stem cells by using the eggs of other animals, such as cows, together with the genes from a human. The nucleus from a human cell, just a connective-tissue cell from under the skin, could be transplanted into a cow's egg and then embryonic stem cells isolated at an early stage of embryonic development. All the key genes would be human and only a few genes that code for proteins in the mitochondria would be those of the cow. The isolated cells could be very valuable material for studying human genetic diseases, as the nucleus would be taken from an appropriate patient. The UK authorities are quite happy with such studies, though some religious groups have protested.

Many of the strongly argued ethical objections to obtaining human embryonic stem cells from the early embryo are based on the grounds that the embryo may be damaged or die – which, it is claimed, is effectively killing a human and showing a total lack of respect for human life. What is critical for some in this context is that an embryo is actually created for research or therapeutic purposes: this raises a wider range of objections, in that a potential life is apparently created for a specific purpose.

The Roman Catholic Church has taken the view that the fertilised egg is equivalent to a newborn baby or an adult. No scientific reason has been given for this view. Even Thomas Aquinas argued that an embryo or foetus is not a human person until its body is informed by a rational soul. St Augustine put 'animation' – the beginning of life, the time when the soul entered and the embryo/foetus could be baptised – at 40 days for males and 80 for females, but later changed it to 40 for both. In the last thirty to forty years the Catholic Church has decided the soul was there from the very beginning – despite there being nothing in the Bible to support the view that the early embryo is already a human being. The Catholic Church and many evangelical Protestant groups have called for a ban on all embryonic stem-cell research, saying it is an assault on innocent human life. Many other Christian churches and Jewish groups, on the other hand, favour embryonic stem-cell research, pointing to potential cures for serious medical conditions.

In the USA the National Institute of Health has removed the word 'embryonic' from its register of stem cells and replaced it with 'pluripotent'. There are even objections to using human embryos for research. But the early embryo is not a human being from the moment of fertilisation, and thus should not be used as an argument by pro-lifers. For example, the early embryo may develop into twins at two weeks of age, and so is clearly not already a human being. The Human Embryology and Fertilisation Authority in the UK is responsible for enforcing the law that no experiments are carried out on embryos after fourteen days without their permission. Failure to obey the law can lead to criminal charges.

About a quarter of all fertilised embryos die early on in the mother's womb, and in the UK some 200,000 abortions are performed each year. Can one really believe that their death is in any way similar to the death of a baby or an adult? My own view is that the embryo becomes a human being only when it can survive outside the mother with minimal technical support, that is at about 36 weeks. Anne McLaren, a leading worker in this field, wondered how many of those who believe the fertilised egg to be human would choose to save a hundred fertilised eggs or early embryos in a burning building, rather than a single baby.

It is also essential to realise that human reproduction by IVF involves the fertilisation of eggs which are allowed to develop to a stage when one or more can be transplanted into the mother, and that in this process large numbers of such embryos are effectively destroyed as they are not transplanted. In the UK between 1991 and 2005 over one million embryos created by IVF were never used. There is no ethical difference between IVF and creating stem cells, as both require the creation of embryos which may later die or be disposed of. One can be, for religious reasons, against both, but not for one and against the other. IVF has been of enormous value, and so too will be stem cells.

There are undoubtedly those who have a sense of revulsion about taking early human embryos and using them to obtain cells for implantation or using them for tissue engineering. But the negative response is not a good way to judge such issues. To whom is harm being done? People were horrified by the first heart transplants, but now there are thousands of grateful recipients. When someone with such views needs to have their nerve cells or blood-forming cells replaced, their negative response may turn to a grateful thanks.

To place the nature of the human embryo in a clearer light, we now turn to another unique aspect of life, the development of all animals from a single cell, the egg.

7

How We Become Human

how we develop from a single cell

It is still amazing, and to our ancestors would have been totally unexpected, that just a single cell – the fertilised egg – very reliably gives rise to something as complex as we humans, or as big as an elephant or as small as a fly. How does it do it? It has been known for centuries that we and other animals develop from an embryo. The problem was the nature of that embryo and how development occurred. There could be no progress until it was known that we are a set of cells and come from a single cell, the egg, when it is fertilised.

Hippocrates, in the fifth century BC, tried to understand our origins in terms of fire and humidity, together with wetness and solidification. Then a century later Aristotle raised questions that took centuries to answer. He asked whether all the parts of the embryo come into existence at the same time, or whether they appear in succession? Is everything patterned and preformed from the beginning, or is it more like the knitting of a fisherman's net? These two processes came to be known as preformation and epigenesis. Aristotle promoted epigenesis.

Aristotle's influence on how the embryo develops was enormous and, although there was no evidence for or against the idea, his ideas persisted until the seventeeth century. There was then a strong rejection of epigenesis, as it seemed unbelievable that physical or chemical forces could mould the embryo and give rise to forms as complex as humans. Thus the contrary view, that the embryo was preformed from the beginning, held sway. There was a belief, religious in origin, that all embryos had existed from the beginning of the world. Even the brilliant eighteenth-century Italian embryologist Marcello Malpighi could not free himself from the pre-formationist doctrine. While he provided an accurate description of the development of the chick embryo, he remained a preformationist and argued that the early stages were too small to be seen. Other preformationists claimed to be able to see an embryo fully formed in the head of each sperm. When the French champion of pre-formation, Charles Bonnet, was told that if preformation were true then the first rabbit would have had to contain more than a billion billion rabbits, he replied by saying that it was always possible to crush the imagination under the weight of numbers by adding zeros.

But the discovery of cells completely changed ideas about the development of the embryo when it was recog-nised, in the 1840s, that the egg was itself a cell that gave rise to all the cells in the body.

How it actually does so involves complex interactions between cells, and the rearrangement of groups of cells, often into sheets, as the embryo develops. How do the early cells know what to do? One possibility was that the egg was already highly patterned and, as it divided up, it acquired determinants that were distributed in a complex mosaic which determined the fate of the cells. It was a theory a bit like preformation. But then came crucial experiments. One set of experiments supported this view, since if one of the two cells of a frog embryo after the first division was killed with a hot needle, the other cell developed into a well-formed half larva. But then, over one hundred years ago, Hans Driesch separated the cells at the two-cell stage of the sea urchin embryo and observed that each developed into a small but normal larva. And today it is well known that one can get twins developing from the separation of many animal embryos at the two-cell stage. Human identical twins come from the division of the embryo at a much later stage, two weeks after fertilisation.

Our own development begins after fertilisation, and there is little evidence for any patterning in the egg itself. After fertilisation, cell division then takes place without growth and results in a group of some 30 cells inside a layer of cells that form a hollow sphere. The outer layer of cells will not contribute to the embryo proper, but to other structures like the placenta. It is the inner cells from which we develop and from which embryonic stem cells are derived. They can give rise to any type of body cell. This makes sense, as at this early stage these inner cells give rise to everything, and there is no difference between them at this stage in development. The interesting question is how the cells then know where to go and what to do. How do they know whether to be a part of an eye or the stomach? It all is due to the network of gene and protein interactions which determine that the right proteins are made at the right time and place and the cells do the right thing in order for the embryo to develop.

The development of the embryo involves five main cellular mechanisms: cell division; pattern formation; change in form; cell differentiation; and growth. Early cell division in the embryo is known as cleavage; at this stage the egg is divided up into a number of smaller cells, some of which are the embryonic stem cells. Pattern formation is the development of the spatial organisation of many structures like limbs and the nervous system. Pattern formation involves a process whereby cells are assigned identities so that they will develop a pattern such as that, for example, seen in the arm, with its large shoulder bone at one end and fingers at the other. If pattern formation is like painting, change in form is like sculpture, as the cells undergo considerable movement and sheets of cells change their shape. Our spinal cord is initially a flat sheet of cells; this curls up, then the curled up edges meet and fuse to create a hollow canal, and it becomes the spinal cord. Of course there are many nerves in that cord, and how they form is cell differentiation. Cell differentiation gives rise to the several hundred different cell types in our bodies like muscle gut, skin and so on. Then there is growth in size of the different parts of the embryo, most of it after we are born.

All these activities are controlled by the types of proteins that are present in the cells, and these are determined by which genes are active. Turning these genes on and off is thus a fundamental process of development. The genes in the cells of the developing nose are the same as those in the developing toe, but the genes that are active in these two regions are quite different. We have already seen that there are control regions for genes to which transcription proteins bind and thus turn a gene on or off. Changes in gene activity are determined both by the sequence of events in a cell, and by signalling between cells. For example, if gene Z is activated by an external signal, then its protein could be a positive transcription factor for gene Y and a negative factor for gene A, and these in turn could result in further genes being turned on and off. Several thousand genes are active in each of the cells in the embryo that require control during embryonic development. There are also many housekeeping genes whose function is to maintain the cells' day-to-day activities; many of these are common to all the cells, and are not part of the embryo's developmental programme.

All the cells that give rise to you and me come from a sheet of cells that formed from those early inner cells. Building a human body from that sheet of cells is, even for the clever cells, a very complex task. There has not only to be patterning so that the cells know what to do at different places and times, but also major changes in shape and movement. The three germ layers that make up our body come from this sheet: the ectoderm, mesoderm and endoderm. The ectoderm is our outermost layer and forms the skin and nervous system; the mesoderm beneath it contains muscle, skeleton, heart, blood and kidney. The third and innermost layer is the endoderm; this contains the gut, liver and lungs. All of the future three germ layers come from the original sheet, and so the endoderm and mesoderm must now move beneath it, leaving the ectoderm on the outside of the sheet. This is a process known as gastrulation.

It is a peculiarity of animal embryos that the cells that will form the inner structures like the gut and the muscles, skeleton and heart, are initially on the outside of the embryo and must move inside during gastrulation. Early patterning of this sheet specifies the main axes of the embryo, like which end is the head and which the tail, and also specifies the three germ layers, and the mesoderm and endoderm then have to move beneath the sheet to their proper position. I am known for my statement 'It is not birth, marriage or death that is truly the most important time in your life, but gastrulation.' It has some truth.

Gastrulation in human development is quite complex and difficult to describe, but the essential features of the process can be nicely illustrated by gastrulation in the sea urchin embryo. Gastrulation is a dramatic example of a change in the form of the embryo, and the cells must provide forces for the movements. Sea urchin eggs and embryos are wonderful for observing cell behaviour during gastrulation as they are both easy to obtain and transparent. One has simply to open the female urchins in the mating season and wash the eggs into a dish; then one puts in a drop of sperm from a male and fertilisation occurs. Within an hour each egg begins to divide, and it does so about ten times over the next ten hours to give a hollow ball formed by a sheet of cells just one cell thick.

The cells then undergo gastrulation and form a feeding larva. The first stage of gastrulation in sea urchins is the movement into the hollow interior of the embryo of about sixty cells from the region where the gut will form. These cells give rise to the mesoderm, the embryo's muscle and the cells that will lay down the skeleton. They find their way to the right place by putting out long, fine, finger-like extensions called filopodia, which can contract and pull the cell forward when they make firm contact with a cell at their tip. It is like climbing out of the sea on to a slippery rock – you come out where you get the best grip. Then the gut – the endoderm – forms by an infolding of cells from the same region, and this infolding extends right across the hollow interior till it meets the cells on the other side and forms a mouth. It is as if you have a balloon and push your finger into it so far that you touch the opposite side of the balloon. In the embryo the tip of the gut sends out filopodia which pull the gut across to the other side. When it touches the other side in the embryo it fuses with it and an opening forms: the mouth. It is an amazing process to watch on film, and shows it is possible to account for all the changes in the shape of the embryo in terms of quite simple cellular forces consisting mainly of extension and contraction. Similar processes were involved when we were gastrulating, but the system is much more complex. Of course all these activities are due to proteins being made at the right time and place due to genes being turned on and off.

The first clear sign of gastrulation in humans is the development of a dense streak of cells from the future rear end of the sheet, which moves forward and then forms a furrow which is related to our main body axis. At the tip of this streak is an organising region known as the node, and if the node is grafted to another edge of the sheet, a new streak will develop. As the streak moves forward cells from the side move towards it and then through the streak and into the region beneath the sheet to give rise first to the endoderm and then the mesoderm. When all this inward movement is complete, the streak begins to move back towards its origin at the rear of the embryo. The future head region is now at the tip of the streak.

As it moves back, the surface layer, the ectoderm, begins to fold in the midline to form a tube that will give rise to the spinal cord and brain. Beneath it, the retreating mesoderm leaves behind pairs of small blocks of tissue, one on each side, known as somites, which will later give rise to your vertebrae and the muscles of your back and limbs. The formation of this segmental pattern of somites down the embryo is determined by an internal clock in the cells that give rise to them – the oscillations of the clock are converted into separated blocks of tissue as the mesoderm retreats. The cells form a somite each time the oscillation is at the right stage and they are at the right position at the tip of the retreating mesoderm. The essential body-plan of the embryo is now apparent, and the limbs will develop later from buds that grow out from the main body.

While the head/tail and front/back of the embryo can now be identified, there is still the problem of left- or right-handedness. Several of our body organs are symmetrical, like our limbs, but our heart is on the left, the liver has just a left lobe, and the stomach and spleen lie to the left. Handedness is subtle. Could you tell by a phone conversation with a symmetrical creature in a dark room which side of their body is left and which right? No, you could not – it is something that has to be shown. Also, why does your left hand look like a right hand when you look in the mirror? The answer is that left and right are by convention related to the front/back and head/tail of the body, and so when front/back is reversed when you look in a mirror, handedness is also reversed. The way left is specified from right in the embryo is by cilia in the node at the tip of the streak moving fluid from the right side to the left; a gene on the left side is turned on and sets up a series of interactions that makes the left side have genes turned on that differ from the right side. About one in 10,000 people have their asymmetry reversed. They are otherwise quite normal, although some may suffer from respiratory problems.

All this development is due to cell activities and is controlled by genes and the proteins they code for. It is essential to realise that the genes do not provide a blueprint for the adult into which the embryo will develop. Rather the genes in the developing embryo provide the basis for a generative programme, one state determining the next. A good analogy is origami, the art of paper-folding. It is quite easy, for example, with the correct successive folds to end up with a relatively intricate bird-like form, but this form cannot easily be inferred from the foldings, each of which has complex consequences on the effect of the next fold. So it is with gene action – each state in development is determined by the previous one. Development is, like origami, dependent on a generative programme. It is also worth noting that the complexity of development is due to the complexity of individual cell behaviour rather than the complexity of interactions between the cells of the embryo. Signals between cells do not pass on complex information, but essentially select from one of the current states that the cell could enter. It is the complexity of the cell's internal state that really matters.

Communication between the cells in the developing embryo and other nearby cells helps turn on genes at the right time and place so that the proteins can do their special work. Like instructions for origami about where and when to fold the paper, genes are made active where and when particular proteins need to be made. The signal the cell receives rarely enters the cell; instead it almost always binds to the surface membrane, setting off a series of interactions within the cell which often leads to activation or inactivation of genes. It is thus a bit like pressing the button on a juke-box to get a particular tune. The complexity comes from the sequence of interactions within the cell that result from a signal arriving at the cell membrane.

In many situations, and particularly in the development of the embryo, signals that arrive at the cell membrane lead to a particular gene being turned on. This signal at the membrane is often relayed via a cascade of protein interactions, which can involve adding a phosphate to a variety of proteins, from the membrane to the cell nucleus, and can activate or inhibit gene activity. This complex relay process, known as signal transduction, has been likened to a Rube Goldberg cartoon in which a man has a relay mechanism for raising his umbrella when it rains: the rain first causes a prune to expand and so light a lighter which starts a fire which boils a kettle which whistles and frightens a monkey who jumps onto a swing which cuts a cord releasing birds who, when they fly out, raise the umbrella.

The cartoon illustrates an important feature of the signals that cells send to each other. Cells rarely pass on direct messages in the sense of giving neighbouring cells new information – there are no letters or notes being delivered. One might have thought that protein messages that were passed from one cell to another would enter the receiving cell, but this is not the case. Protein signals activate receptors that send signals into the cell which involve protein and protein complexes interacting. The conversations between cells depend on quite a small number of molecules, mainly proteins, which are used again and again, and the response reflects the current state of the cell, which can depend on its history. So, following the concept in the cartoon, rain on the prune could lead a to different result, perhaps the man letting a dog out. Only a few signals do enter the cell and these are usually hormones. The Goldberg cartoon illustrates this nicely.

Many of the key genes controlling our development, as well as that of other vertebrates, were originally identified in a key model organism, the fruit fly *Drosophila*. These flies were used to discover many aspects of genetics and how genes and proteins control cell behaviour. (The flies probably often comment on how surprised they are that their mechanisms of development are so similar to those of humans.) One way the similarities were discovered was by taking genes that were playing an important role in the development of the fly, and then looking for a related gene in vertebrates like the mouse or chick. A nice example is Sonic hedgehog. This gene comes from a gene in the fly called hedgehog, as its mutation gave the surface of the fly larva a hedgehoggy appearance. When the related gene was identified in vertebrates, the scientist called the gene Sonic hedgehog after one of his son's computer games that had a related name. (Developmental biologists have the right to name genes that they have discovered which control development, and they enjoy choosing the name.) Sonic hedgehog is a signalling protein used again and again in our development – the cells respond differently to the signal because they have a different history and so contain different proteins. Other model organisms like frogs and chickens have also made major contributions to our understanding of development. The basic mechanisms for development are similar in most animals; evolution has used again and again those mechanisms that have been successful to generate variety.

In development communication between cells is fundamental in determining how cells will develop. We will now look at how pattern formation can be illustrated, and even begin to be understood, in terms of the French flag problem. The French flag has a simple pattern, one third blue, one third white and one third red, along just one axis. The flag can be of any size but the pattern is always the same, and this can be thought of as similar to those embryos which are able to develop the same form even if initially they are different sizes – consider identical twins that develop from a single egg, each having half the number of cells that are normally involved in a single embryo. Given a line of cells each of which can differentiate into a blue, red or white cell, and that the number of cells in the line is variable, what mechanism could reliably ensure that they form a French flag pattern?

Imagine that you are in a line of people and each of you has three pieces of paper, one blue, one white and one red. You are to be visited by some famous person from France, and it is required that at a signal your group looks like a French flag, with each of you holding up the appropriate piece of coloured paper. Of course some of your colleagues think this ridiculous and leave the line just before the arrival of the guest. But if you know your position in the line, then you would also know which piece of coloured paper to hold up, and this requires continuous signalling along the line. So it is in the embryo.

The biological solution is based on the cells acquiring information so that each cell has a positional value which specifies its place in the line with respect to the boundaries at either end. After the cells have acquired their positional values they interpret them by differentiating according to their genetic programme: those in the left-hand third become blue, those in the middle third white, and those in the right-hand third become red. A virtue of this mechanism is that many different patterns can develop using the same positional values, with different interpretation giving the required pattern. Moreover it is easily extended to a two-dimensional system with positional values being specified along two axes. We shall see examples of just this.

How could position with reference to the ends be specified? Not a simple problem. Because all patterns are initially specified in small groups of cells, the line in any direction is always smaller than about thirty cells, and this persuaded Francis Crick to propose that the diffusion of a chemical from one end could specify a position by the cells reading the local concentration. If there were a source of a diffusible substance at one end and a sink at the other, so that the concentrations at the ends were kept constant, the diffusion of the signal would result in a straight line of graded concentration. Then, with specific thresholds, a French flag would develop – above a particular concentration blue cells would differentiate, while below a much lower threshold only red would form, while white would develop between them. Attractive as this is, it now seems that diffusion is too unreliable a mechanism to give cells their positional information. Just how cells get their positional values is still controversial, but probably involves the cells talking directly to each other by an exchange of information at cell junctions.

Very good evidence that cells do have positional values comes from studies on regeneration. Several types of frogs can regenerate their limbs, and this requires that the cells have positional values along the limb so that regeneration starts from the cut surface and only positional values towards the tips of the fingers are specified. A protein on the membrane of limb cells has been identified and it is graded from high to low from the frog's shoulder to the fingers. It is also possible to alter positional values along the limb by treating the regenerating limb with retinoic acid, which makes the cells have higher values of the protein, and so become like those cells at the shoulder. If the hand is amputated, then just a hand is regenerated. But if retinoic acid is added during regeneration the regenerating cells think they are at the shoulder, and a whole new limb regenerates from the hand region.

Another example of patterning involving positional values comes from *Drosophila* and illustrates that the positional values in the leg and antenna of the fly are the same. Both are quite long structures, projecting one from the body, the other from the head, but with very different shapes. However a mutation in a single gene can result in the fly developing a leg instead of an antenna projecting from the head. It is possible to express this mutation in a small region of the developing antenna, and then that region develops leg structures corresponding to that position along the leg. Clearly leg and antenna have identical positional values, but interpret them differently because of particular control genes. The many targets of such control genes that determine the interpretation of position have yet to be identified. In addition, the cells in both the developing leg and antenna communicate and interact with each other to specify position. That a single gene can have such a powerful effect on the interpretation of positional values is amazing, and the mechanism still requires further research.

The importance of the control regions of genes in pattern formation comes from another, rather dramatic fly example. Early in fly development some genes are expressed on the dorsal side in seven stripes, which can be seen if the protein they code for is labelled. It was originally thought that the stripes were specified by an underlying wavelike pattern in some substance, the stripes developing at the peaks of each wave. This is not the mechanism, but nor is it one based on positional information. Each stripe is specified separately from all the others. The way this happens is due to the stripe gene having seven different control regions, resulting in it being expressed in the seven stripes. Each of the control regions is activated by different proteins synthesised earlier in development, acting as transcription factors present in the region of each stripe. This illustrates a fundamental mechanism for the control of cell behaviour during development – the control regions are at the core.

A good example of pattern formation in our own development is the limb. Studies on the chick limb have been particularly helpful as the chick embryo makes it quite easy to open the egg and operate on the developing limb. Also helpful is the mouse, where the genetic control of limb development has been studied. The limb grows out from the body when the main body is already quite well developed. The early limb bud is like a flattened balloon with a sheet of cells as a covering, and inside are the dividing cells which blow it up and cause it to grow out. Later the mass of cells inside will give rise to the cartilaginous elements, the precursors of our bones. At the tip of the sheet of cells is a thickened ridge rather like that on a hot water bottle, and this gives the limb the somewhat flattened form which you can see by looking at your hand. This ridge secretes a protein that specifies the cells beneath it as the progress zone. It has been proposed that patterning occurs in the cells in the progress zone, as this is where the cells get their positional values.

At the edge of the progress zone on the side, where your little finger will develop, is a special signalling region that secretes the protein Sonic hedgehog. In the developing limb, Sonic hedgehog is involved in specifying positional information along the axis which runs from the thumb to the little finger. The concentration of Sonic hedgehog is highest at the 'little-finger' end, and decreases towards the 'thumb' end. Its concentration is thought to specify position, so that the digits from little finger to thumb acquire their identity at positions along this concentration gradient. In the chick there are three digits, known as 4, 3, 2 going from 'little finger' to 'thumb'; the normal pattern is 432, and digit 4 is believed to be specified by a high concentration of Sonic hedgehog while digit 2 is specified by a lower concentration. If a Sonic hedgehog region is taken from one embryo, it can be grafted to the 'thumb' margin of the chick limb bud, and so a mirror-image gradient is set up, the limb widens, and six digits develop, giving a digit pattern 432234. The proposed specification of position by the gradient may also involve cells talking to each other, so that digit identity is reliably specified.

Sonic hedgehog helps control the development of digits in both arms and legs. The response is different in arms and legs due to different genes having been turned on because of their different position along the body axis. So in the leg extra toes are formed when an additional Sonic hedgehog region is grafted. Children are rarely born with extra digits but when this happens it is due to the development of an extra Sonic hedgehog signalling region at the thumb margin of the limb. Programmed cell death (apoptosis) also plays a role in limb development as the development of the digits requires that the cells between them die – if they did not, our hands would become webbed and duck-like. Thus programmed cell death is part of pattern formation and is under genetic control.

For the main axis of the limb, from shoulder to hand, it is thought that the time cells spend in the progress zone could determine their positional value. All the cells in the progress zone under the ridge are dividing, and they may measure how long they remain in the progress zone before being pushed out by the dividing cells. The cells that remain in the progress zone the longest become digits, while those that leave it early on become the next set of bones, the radius and ulna. This model, in which position along this axis may be based on a timing mechanism, may provide an explanation of the effects of thalidomide, which when taken by pregnant mothers resulted in limb deformities in the child. If thalidomide kills cells in the progress zone, then it takes time for the cells to repopulate it, and so it takes more time before a significant number of healthy cells begin to leave it. The result is that only structures in the hand region develop. There is evidence that thalidomide blocks blood vessel development and this could lead to cell death in the progress zone, so that children are born with just a hand at their shoulder.

All the muscle cells in our limbs migrate in from the somites, the small blocks of tissue that lie beside the neural tube in the embryo. Unlike the cells in the limb the cells that will develop into muscle have at an early stage no positional values and are initially all the same, a true democracy. When the muscle precursors enter the limb they are guided to specific sites by the limb cells with their positional values, and then they develop into muscle cells and join up with bones and tendons. They are again truly democratic and indiscriminate, and will join up with any bone or tendon they come into contact with. One can see this in the chick embryo wing if the tip is rotated through 180 degrees – tendons on one side now come into contact with muscles on the other side and join happily with them.

An as yet unknown mechanism is the basis for specifying position along the main body axis and determining where the neck, ribs, limbs, and lower back will develop. This mechanism activates a special set of genes known as the Hox genes, also originally discovered in the fly. The name comes from mutations in these genes which can result in a homeotic transformation, in which one structure replaces another as in the antenna to leg transformation in the fly *Drosophila* discussed above. The fly has one set of these genes expressed along the body in the order they are expressed on the chromosome. This is the only known case of a spatial correspondence between the order of genes on chromosomes and where they are expressed in the embryo. We have four sets of up to thirteen homeotic genes closely linked together on four different chromosomes, due to the duplication of the original set way back in our evolution. They are expressed at different positions along our head-to-tail axis and are essentially the local positional values, determining local development such as where ribs will form and limb buds develop. Genes at one end of the chromosome are expressed at the anterior end of our body, while those at the other end are expressed at a later time in much more posterior regions. For example Hox gene A1 is expressed in our head region and Hox gene A13 at the base of our spine.

The spinal cord is a good example of morphogenesis (change in form) based on mechanical processes that would impress any engineer. The tissue that will give rise to the brain and spinal cord is specified quite early in development and located as a narrow, rather flat sheet that extends backwards on the upper side of the body. Formation of the tube that will give rise to the spinal cord is like folding a sheet of paper. The first step is the formation of a groove along the midline, which results in the regions on either side of the groove rising upwards as folds. These folds move towards the midline and meet and fuse. The result is a tube of just a single layer of cells with a hollow core. These changes in form are due to forces exerted by the cells that form the tube. The formation of the initial groove is caused by the cells in the midline changing their shape to become wedge-shaped, probably due to a contraction beneath the membrane at the cell's inner face. Following fusion, the tube separates from the adjacent tissue; this reflects a change in the adhesion molecules on the cell membranes that keep cells in contact with each other. In the posterior region of the body the tube folds to become a solid rod, and then cells in the middle die to turn it into a tube. Its later development will be considered with the nervous system.

Our face largely forms from a group of cells, the neural crest, which leave the neural tube and give rise to many cell types. Migrating away from the tube, they differentiate into cell types that include pigment cells, the cartilage of the head – my large nose comes from the neural crest – and the sensory nerves in the spinal cord. The pathway taken during their migration is determined by the cells over which they move, as directed by the extracellular material that these cells secrete. These molecules can affect the direction of migration by, for example, the strength of the adhesion between the anterior regions of the migrating cells. If the cell puts out fine extensions at its tip, then it will move in the direction where these make the strongest contacts.

The neural crest cells have the capacity to develop into a variety of different cell types when they leave the neural tube – they are multipotent. As they migrate their developmental potential becomes smaller because of the signals they receive from the neighbouring cells over which they move. Their differentiation depends on where they find themselves: local signals result in the activation and repression of the genes that lead to the differentiation of the specific cell types, and there is much signalling between cells, mainly of the Rube Goldberg type.

Our vascular system, composed of blood vessels, is the first organ to develop in the embryo. The key cell type from which it develops is a cell that lines the whole circulatory system, heart, veins and arteries. They initially assemble into tubes, and one end of each tube is the site of growth where cells proliferate. The cells at the tip of the vessel extend fine contractile processes to extend and guide the growth of the vessel, which is influenced by signals from the matrix through which they are moving. Some of the signalling molecules are the same as those that guide nerves to their destination. These small tubes are specified as arteries or veins at an early stage, even before they start moving, but they can have their identity altered. During their migration, branches develop, again influenced by local signals. Since their function is to bring blood to the different body regions, those regions that will need a blood supply send out signals to attract blood vessels.

The process of cell differentiation, which gives rise to many cell types, is determined by changes in gene expression and the synthesis of different proteins. The transcription of a gene is determined by protein transcription factors binding to the control region. There are some 3,000 different transcription factors in our cells; not all of these are involved in controlling development, some being there to control household genes – those necessary for the day-to-day activities of the cell. Many transcription factors are required to turn on some genes, while only a few may be sufficient for others.

Red blood cells have no nucleus or mitochondria, but are filled with the protein haemoglobin, which makes them red, and it is the haemoglobin which takes hold of oxygen in our lungs, delivers it to our cells, and then takes hold of carbon dioxide which it returns to the lungs. Signalling molecules help to specify which stem cells in the marrow will develop into red blood cells or white cells. Control regions can be far away on the chromosome from the gene they control.

Our haemoglobin is made up of two linked types that join together. These two haemoglobins are on different chromosomes, and are part of two multigene families which code for slightly different haemoglobins. They are made at different times in our development as they bind oxygen to different extents, and there are different requirements for oxygen during our development from embryonic to adult life. One family of haemoglobins is encoded by five slightly different genes, which are expressed at different stages of development; the control regions for these genes are rather complex as they require the switching on and off of the five different globin genes as development progresses. The mechanism controlling their expression makes use of a region on the chromosome a long way away from the globin genes, and this requires bending of the DNA so it can come in contact with the globin transcription complex and turn on the right genes. Once the genes have been transcribed, the nucleus is pushed out of the cell to give a functioning red blood cell.

Our voluntary muscles whose contraction we control – as distinct from those, for example, in our heart and gut – come from the somites, as we have already seen in the case of the limb, and this is also the case with our back muscles. Voluntary muscle cells will develop well in culture. Cells that will give rise to muscle can be isolated from the somites of mice and placed in culture where they proliferate, and only when they stop dividing will they differentiate into muscle. They then begin to synthesise muscle proteins like actin and myosin, and then fuse with each other to form muscle fibres with numerous nuclei. This fusion makes it possible to have large muscle cells.

The development of the nervous system and the connections between nerves will be discussed later, but it is estimated that there may be as many as 10,000 different types of nerve cells in our body. Even though they may look similar, there are subtle molecular differences that give the nerves different thresholds for firing, and different firing patterns.

We now need to look at the germ cells from which we develop.

8

How We Reproduce

how meiosis works

Eggs and sperm are the true royal family in the society of cells. It is only their descendants that will continue to live when all the other billions of cells are dead. In fact the sole role of all those billions in our body is to make sure that the egg and sperm meet a partner and reproduce. That evolution has resulted in a sexual basis for our reproduction, rather than the egg doing it on it own, is a complex story. The sexual basis provides a mechanism for generating more genetic variation as well as eliminating unwanted genes. One mechanism for getting rid of dud genes involves the females choosing their male partner, as in vertebrates, and thus eliminating males with poor genes.

Eggs are the only cells in animals that can give rise to a whole being. Development of the human egg is initiated by the sperm fertilising it and providing it with another set of genes. The development and form of the sperm and egg are quite different, but both are needed as the egg cannot develop independently. Together they can transmit particular human features, from eye colour to the length of the nose, to the next generation. I remain in awe that we come from that tiny cell, the egg, and it is important to understand where and how it develops into the embryo. In many animals there are, in the egg, special proteins in specific regions of its cytoplasm that will later specify development of another egg. In human eggs there are no such elements. So how are eggs and sperm made?

The region that will give rise to human eggs and sperm can be identified about halfway through gastrulation in the mouse and thus probably at a similar stage in humans, and the cells later migrate to the future genital regions, the ovary and the testis. Strange to think that the egg from which we came took this journey from one part of the body to the genital region. The future eggs undergo divisions to increase their number as they migrate, and continue to do so for a short time when they reach the ovary. Then they stop, and there is no further increase in the egg number, which at this early stage is around six million. There are still 23 chromosomes from the mother and 23 from the father in these cells. The future eggs now begin a very special type of cell division, meiosis, which results not only in the reduction of their chromosome number by half, but also a dance by the chromosomes in which the male and female chromosomes, by mixing up genes that come from both mother and father, recombine in different ways to form new and different chromosomes.

In meiosis the future egg cell divides twice but the chromosomes are duplicated just once, so their number is reduced to half; the full number is restored later at fertilisation by the entry of the sperm. The first step in the reduction process of meiosis is the duplication of the chromosomes in the same way as in normal mitosis at cell division. They remain attached to one another and then, cleverly, each pair of paternal chromosomes finds and pairs with the similar paired chromosome that came from the mother. Then the chromosomes exchange some similar regions so that some sets of genes from the mother are now on the father's chromosome, and vice versa. Two cell divisions follow without any further increase in chromosome number, and so give rise to four eggs with a single set of just 23 chromosomes, which have been reshuffled so that each of the four cells has different sets of genes. This provides an almost limitless resource of genetic variation.

The eggs in the ovary enter the first stage of meiosis but never proliferate again, so the total number of possible eggs, some six million, is determined when females are born. But only some 40,000 remain and escape degeneration by the time the female reaches puberty. Each of these has the potential to develop into a child when fertilised. At puberty, the egg cells grow some 1,000-fold in size and continue meiosis. The development of sperm does not involve any meiosis in the embryo; this occurs later in the testis in the sexually mature male. With sperm there is again a reshuffling of the chromosomes. Unlike eggs, sperm continues to be produced throughout the life of the male.

In the development of both eggs and sperm, the final state of their genes must be such that they can give rise to all the cells in the body. But during their development, certain genes in both eggs and sperm are specifically inactivated by a process known as imprinting. Imprinting is a way of blocking the function of a gene, and it is done by methylation – putting a methyl group, a small chemical compound, on to specific regions of the DNA, which stops the genes being transcribed. Some 70 imprinted genes have been identified in mammals but the function of only some of these are understood. The clearest examples relate to the growth of the embryo: a gene coding for a growth factor is turned off in the egg chromosome but remains on in the sperm. The evolutionary explanation is that there is a conflict between mother and father with respect to reproductive strategies. The father wants maximal growth of his offspring, while the mother wants to spread her resources over many offspring and therefore wishes the growth of each embryo to be reasonable but not excessive, and so inactivates the gene that codes for a growth factor.

Certain genetic diseases are related to imprinting. These include Prader-Willi syndrome, which involves a preoccupation with food and learning difficulties, and Angelman syndrome, which results in mental retardation, abnormal gait, speech impairment, seizures, and an inappropriate happy demeanour that includes frequent laughing. The genes involved are on chromosome 15 which are imprinted in the female and there must be paternal expression for normal development. If there is a deletion of these genes in the paternal chromosome, the child will have Prader-Willi syndrome. If the same region is deleted from the female chromosome 15 it causes Angelman syndrome. With Beckwith-Wiedemann syndrome very large babies develop due to an imprinting error that results in the failure to turn off one of the genes that codes for a protein growth factor, so that both maternal and paternal genes are active and there is excessive growth.

Fertilisation, the fusion of the two sex cells, sperm and egg, starts the egg developing. During human sexual intercourse some 300 million sperm are injected into the female vagina, but fewer than a thousand manage to migrate to the waiting egg. They can survive, however, for three days. The sperm swims towards the eggs by waving its tail, but only one of the thousands searching for the egg actually fuses with it, burrowing its way through the egg's protective layer. Once one sperm fuses, there is a block to further sperm entry. Fertilisation causes a release of calcium, and this activates an enzyme that adds phosphate groups to proteins and initiates development of the egg, which now divides and completes meiosis by producing a tiny daughter cell which disintegrates.

For some parents who are having difficulty reproducing, it is now possible to fertilise eggs outside the mother and then transfer the egg back into the mother – the procedure known as IVF, in vitro fertilisation. Egg production is hormonally induced in the mother and then the eggs are removed from the ovary and placed in a fluid medium in a culture dish. Sperm are added to the fluid medium, fertilisation can take place, and the fertilised egg can then be transferred to the mother's uterus. When there are few good sperm available, it is even possible to fertilise the egg by injecting a single good sperm directly into the egg.

Only the nucleus of the sperm normally enters the egg at fertilisation, and after about twelve hours the 23 chromosomes from the mother now join in a nucleus with the 23 from the father. There is now a fertilised egg that can, if all goes well, develop into a human being. Cell division begins.

Our sex is determined by which sex chromosomes the sperm provides. There are two chromosomes linked with sex, X and Y. Males have both an X and Y, and are designated XY, while females have two X chromosomes, and are designated XX. At meiosis during sperm development when the number of chromosomes is halved, half of the sperm carry an X, while the other half carry a Y. Every egg, by contrast, has an X, so it is the Y chromosome that determines male development. If the sperm with an X chromosome enters the egg the embryo is then XX and develops as a female, whereas if it provides a Y chromosome the child will be male. Sexual abnormalities occur if the egg has a chromosome constitution XXY, or just one X. Being a male or female is thus determined by the male sperm at the time of fertilisation.

The Y chromosome determines male development because it contains a gene that codes for a protein that leads to the development of a testis and the production of the male hormone testosterone, which results in the differences between males and females. At an early stage of development the structure of the embryos of human males and females are indistinguishable. Being female is the default state; maleness results mainly from the action of the hormone testosterone, which gives many organs a male form. This is a result of the action of that single gene on the Y chromosome, which causes the embryo to develop a testis rather than an ovary. The testis exerts its influence through the production of the hormone testosterone, which, for example, causes a penis to develop rather than a clitoris in the genital region, although at an early stage there are no differences in the tissue that will give rise to these two structures. It also prevents large breasts developing in the male and has effects on the brain. If males have a rare disorder that makes them insensitive to testosterone, they will have a testis but look like a female. In rare cases an XY individual can be female due to the region of the Y chromosome carrying the gene that is necessary for the development of a testis being lost. There are also cases of XX individuals being male due to part of the Y chromosome being transferred to one of the X chromosomes.

An embryo whose cells are XX and that will develop into a female initially has twice the number of X genes compared to the XY male. This has to be corrected, since there must be the same level of expression of genes on the X chromosome in both sexes. We, like other mammals, correct this imbalance by randomly inactivating one of the X chromosomes in females in each cell early in development. As this inactivation is random we end up with a similar number of male- and female-derived X chromosomes throughout the body. When things go wrong with the number or behaviour of the sex chromosomes, there are abnormalities in development which are considered later in Chapter 13.

The development of eggs and sperm should make it clear why we cannot inherit acquired characteristics like knowledge, memories or skills. A theory proposed in the early nineteenth century by Jean-Baptiste Lamarck to explain evolution suggested otherwise – but how could such activities, largely occurring in the brain, possibly be transferred to the egg or sperm and so be inherited? This cannot happen. However, if certain events, such as an illness or an environmental agent, involve chemicals changing the same genes in both the adult and the egg or sperm that lead to a particular behaviour, then what looks like such Lamarckian inheritance could result.

*

How reversible and plastic are the changes in behaviour of the cells and genes during the development that we have been considering? It was originally thought that genes that had functioned during early stages of development were eliminated and lost. The dramatic experiment to test for such a possible loss, as well as for finding out if the pattern of gene activity in an adult cell could be changed, involved cloning. Could the pattern of gene activity in differentiated cells be made to revert to that found in the egg? Were any genes lost during development?

The experiment to test this involved inserting the nucleus of a differentiated cell into an egg whose nucleus had been removed, to see if the egg could then develop normally. Frog embryos were used, as their nucleus is just beneath the upper surface of the egg and easily destroyed by X-irradiation. The nucleus of a cell, like that from the gut, was then inserted with a fine pipette into the egg, whose nucleus had been destroyed. This resulted in the egg going through early development at least to the tadpole stage. No genes were lost in the differentiated cell. Injection of nuclei from several differentiated cells such as skin resulted in similar development, and this clearly showed that the gene activities in the differentiated cells could be changed back to those which were present in the nucleus of the egg. Moreover the tadpole had the identical genetic constitution to that of the animal from which the differentiated cell had been taken. It had given rise to a collection of genetically identical cells, and this is what is meant by cloning.

Most of the clones that developed from the frog egg did not develop further than the tadpole stage, for reasons that are not known. However if the nucleus was taken from a cell of the early developing frog embryo, normal adult frogs developed. With mammals like mice and sheep, the situation is more complex. A sheep, Dolly, became famous for being the first mammal to be cloned from an adult cell, in this case from the udder, but Dolly died young and with health problems. Moreover, hundreds of embryos from other cloning experiments have failed to develop normally. Cloned mice, cows and sheep have, however, developed into adults. Mice have even been cloned using nuclei from some nerves in the brain, but most nuclei from nerves resulted in the embryo ceasing to develop at an early stage. In general terms, the success of cloning from differentiated cells in mammals is very low, and the animals that do develop often have abnormalities ranging from early death to limb deformities. Moreover, in cloned mice as many as 5 per cent of genes are not correctly expressed. The egg cytoplasm clearly cannot reverse all the changes that took place in the nucleus during its development. No case of successful cloning of a primate such as chimpanzees has been reported, but cloned embryos have been made. All this indicates that cloning of a human is a very dangerous procedure, as the child will most likely suffer from abnormalities. It is therefore right that cloning of a human should be banned – not for ethical, but for health reasons.

Why should anyone wish to clone a human? They may wish to make an individual identical to themselves or to one of their children or to some famous person. But the cloned individual, though genetically identical to the donor of the DNA, will have a different character depending on how they are treated when growing up. There are no good reasons for cloning another human.

However cloning that leads to just an embryo could be helpful for regenerative medicine, as the cells could be used to help a patient without immune rejection.

There have been from the beginning strong ethical objections to the cloning of a human being. Jeremy Rifkin, President of the Foundation on Economic Trends based in Washington DC, demanded a worldwide ban and suggests that it should carry a penalty 'on a par with rape, child abuse and murder'. Many others, national leaders included, have joined in that chorus of horror. There are fears that the child would suffer from being genetically identical to the donor of the nucleus, as the similarities would constantly be examined and it could have unreasonable expectations imposed on it. There is also the fear that a whole number of identical individuals whose donor was criminal could be cloned. Religious opposition is based on objections to interfering with the early embryo, on the grounds – as discussed earlier – that the embryo could already be considered a person. Given all the cell activities involved in our development, is it reasonable to claim that the fertilised egg is a human being and should be treated as such? It seems hard to think of a single cell, the fertilised egg, as a human being. Just consider how much it still has to go through before a human begins to emerge around nine months later.

I have offered as a prize a bottle of champagne to any-one who could show me that cloning a human being rais-es any new ethical issues. That the individual would look like the person from whom the nucleus was taken is of little concern, as many children look like their parents. In addition there is quite often sperm donation from some-one not related to the family. Many parents have unreasonable expectations of their children. No one has claimed my prize, as I added that if I show that they are wrong they have to give me two bottles. It is the almost certain abnormal development of a cloned child that requires human reproductive cloning to be banned, not ethical considerations.

There is in the future the possibility of modifying the genetic composition of the egg in order to give a child special characteristics such as increased intelligence or sporting ability. This does raise ethical issues, as well as the danger that such interference could have unanticipated bad effects on the child's development. Eggs and embryos are highly complex and we do not yet understand them sufficiently to introduce such changes with confidence.

Lewis Thomas, a highly regarded physician, wrote in his book *The Medusa and the Snail* about the 'miracle' of how one sperm cell fuses with one egg cell to produce the cell we know as a zygote, which, nine months later, will become a newborn human being. He concluded: 'The mere existence of that cell should be one of the greatest astonishments of the earth. People ought to be walking around all day, all through their waking hours, calling to each other in endless wonderment, talking of nothing except that cell.'

9

How We Move, Think and Feel

how nerve cells communicate

The grey, soft, lumpy mass that is the human brain is rightly regarded as probably the most complex structure in the universe. There are some 100 billion nerve cells, also known as neurons, in our brains, and even more supporting cells, and that this mass of cells does all our thinking and feeling is almost beyond belief. Everything we do is determined by this impossibly complex society of nerve cells. But exactly how they communicate in their complex networks so we can think and feel, and be conscious of what we are doing, and can walk, if we wish, in a straight line while having creative thoughts at the same time, is a puzzle.

It is all due to signalling between the nerve cells. Long extensions from nerve cells are often bound together in a common cable-like bundle – this is what is commonly called a nerve and can be very long. In addition to nerve cells there are even more numerous supporting cells in our brains known as glia. Their function is not to signal but to provide insulation, support, removal of debris, and also to nourish nerve cells.

Nerve cells both communicate with other nerve cells in the brain and also send out signals via long nerves that cause muscles to contract. One of the main functions of our brain, and its evolutionary origin, is to control the contraction of muscles, and so how we move. Given the complexity of the connections in the brain, how is this set up during its development? A nice question is the extent to which different nerve cells have unique identities that are specified during embryonic development. In fact different combinations of just 37 genes could, in principle, specify every one of the billions of cells in the brain. But how this enormous society of brain cells talking to each other can give rise to all our thoughts, emotions, movements and even consciousness still remains a mystery.

Nerve cells have the prime duty to carry messages rapidly to other cells, mainly to other nerve cells but also to muscle. Just as important are those nerves that carry information from different parts of the body to the brain, and which are the basis of how we sense what is going on around us. There are nerves involved in our sense of touch, nerves that enable us to feel pain and thus try to remove the cause, nerves in our eye that carry the information that enables us to see, and nerves for detecting smells and the local temperature. All these transfers of messages involve an ingenious way of sending a message along the nerve cell based on changes in the electrical charge across the cell membrane.

Every nerve cell has a small region similar in size to most other cells containing the nucleus with the genes and mitochondria; this is where protein synthesis takes place, and from this region there are long, thin extensions. Usually there is one particularly long extension known as the axon, which will take signals to other nerve cells and to quite distant targets like muscles. The axons from many nerve cells are often bundled together in what we usually refer to as nerves. The axon may be just a hundredth of a millimetre in diameter, but can be more than a metre long, as is the case of nerves from our spinal cord to muscles in our arms and legs. The axon can branch near its end and so signal to many other nerves and muscles at the same time. The axon may be covered by special fatty insulating cells – Schwann cells – that make the signal travel faster down the nerve. (The loss of these insulating cells is the cause of multiple sclerosis.) There are also smaller extensions from the region where the nucleus is located known as dendrites, and their function is to receive signals from other nerves. In the brain there can be very extensive inputs to a nerve cell by these dendrites, as many as 100,000 of them for a single cell. Contact regions between nerve cells where they communicate with each other are called synapses, and each nerve cell has between 5,000 and 200,000 synapses. All these connections between the vast number of nerve cells makes our brain incredibly complex, and to fully understand how it functions will take much new work and probably a long time.

All nerve cells convey signals along their axons and dendrites in the same way, by changes in the electrical charge difference – the electrical potential – across the cell membrane. In the resting nerve cell, sodium ions are being continually pumped out, and the concentration of sodium is thus higher outside the cell. The result is a negative electrical charge on the inside of the cell membrane compared to the outside. If the membrane of the nerve cell is altered by a stimulus from another cell so that it has a less negative value, then pores in the nerve cell membrane open up and allow sodium to flow in, and this causes the electrical potential to change from negative to positive. Not liking all those sodium ions flooding in, the cell quickly closes the open channels and starts to pump out sodium again. But these local changes in electrical potential have further key effects, which are the basis of nerve cell signalling. The local change in potential from negative to positive depolarises the neighbouring region of the membrane, which then goes through the same process of allowing sodium in and changing the local membrane potential. It is rather like the nerve passing down a message saying 'positive'. This message, known as an action potential, moves along the axon and is the way nerve cells send a signal, or what is commonly called a nerve impulse. There are billions of nerve impulses running down your billions of nerve cells right now.

When a nerve impulse reaches the end of the axon it arrives, if there is another nerve at its end, at a synapse. Here, between the end of the axon and the membrane of the receiving cell there is a narrow gap which the electrical signal cannot cross, so the nerve sends on a different type of signal to communicate with the responding nerve cell at the synapse. The arrival of the impulse at the synapse causes the release of a small amount of chemical neurotransmitters, which then diffuse across the synapse gap and bind to receptors on the membrane of the responding nerve cell. The neurotransmitters like adrenaline, dopamine and serotonin are contained in small vesicles, tiny membrane-bound structures, which then fuse with the membrane. They can encourage the responding nerve to send a similar impulse by locally altering the electrical potential by making it more positive, or inhibit the cell from doing so by making it more negative – depending on the type of cell and the neuro-transmitter. The electrical impulse is thus converted into a chemical message, which is then converted again into an electrical one.

These interactions at synapses are the basis of nerve-cell communication in our brain, as they determine which cells will conduct an impulse. A nerve cell can have thousands of synapses on it with inputs from other nerve cells, and together they determine whether or not the nerve cell will fire and conduct an impulse.

Muscle cells are instructed to contract by a similar method. All the movements which we consciously make are due to the contraction of our skeletal muscles. These muscle cells can be very large compared to our other cells, and have a fibrous shape with a length of several centimetres. They are not simple cells, as pointed out earlier, but a combination of many cells that have fused together, and so have many nuclei. Our other muscle cells, like those in the heart, are small single cells, but all contract because of the sliding of the proteins actin and myosin past each other as described earlier. The initiation of a contraction is due to the chemical released at the special junction, like a synapse, at the end of the muscle nerve. It causes a change in the electrical charge in the muscle cell membrane. This in turn releases calcium inside the muscle cell, leading to a change in the myosin so that it interacts with actin, and results in contraction. That is the basis of how we raise our arm.

A colleague of mine, Geoffrey Burnstock, wanted to know how nerves activated contraction in involuntary smooth muscle like the gut. It was assumed that the process was similar to that for voluntary muscle, like that in our arms and legs: an electrical impulse went down the nerve and then released a neurotransmitter, which bound to the muscle membrane and activated it. He developed a method for recording the electrical activity of smooth muscle, and found that blocking all the known neurotransmitters did not stop the nerve activating the muscle. Some unknown transmitter had to be involved. In order to identify the neurotransmitter he tried many different substances and, in 1970, made the fundamental discovery that ATP, a universal molecule used by cells for energy, was the transmitter. In 1972 he coined the term purinergic signalling. Few believed him and his theory was ridiculed, but he turned out to be right. These discoveries have important implications for treating a number of medical conditions including pain, migraine, cystic fibrosis, bladder incontinence and cancer.

A major function of nerve cells in our bodies is the processing of information from both outside and from within our bodies, and then signalling to the muscles to give an appropriate motor response. All this is done by specific sets of interconnected nerves in the brain and spinal cord. A nice simple example is provided by the stretch reflex, the knee jerk – we have all had the doctor tap the tendon running down the outside of our knee bone, and seen our limb kick up. The tap causes a transient stretch of the extensor muscles of the leg, which produces a signal to the nerves that control contraction of the muscle. There are also inhibitory signals to the muscles that would oppose the knee jerk, and several hundred nerves are involved, so even this simple example has its complications.

*

The interactions between nerve cells are more complex than the interaction between nerves and muscles. Muscles have only one nerve attached to them, a motor nerve, and they only receive activating inputs, although the stimulus can vary in strength. By contrast, central nerve cells receive both activating and inhibitory inputs. A nerve cell in the spinal cord receives hundreds, even thousands of connections via synapses from other nerve cells. There are a variety of transmitters in nerve-cell communication which can stimulate or inhibit the receiving nerve cell, and whether the receiving nerve cell generates a nerve impulse depends on the balance between the activating and inhibitory signals it receives. As many as 50 activating impulses may have to be received in order to overcome negative signals and make the nerve cell fire.

Then there are special sensory nerve endings in our skin that are sensitive to their environment; these can respond to pressure, heat or pain and transmit the information to our brain. One set is sensitive to touch, having a specialised end surrounding the nerve terminal. Our finger tips are the most sensitive. Hair cells also respond to touch. Activating the sensory nerves in any specific region of our skin leads results in a map of the body being plotted on the brain: for example, touching the genitals leads to response at one end of the map in our brain, while touching the tongue is relayed to the other end. In a famous series of experiments on anaesthetised patients, Wilder Penfield stimulated various regions of the brain that are involved in feeling sensations. This activation of the brain produced in the patient a feeling that a particular part of their body was being touched.

Penfield was thus able to map on the brain the regions of the body that responded to touch.

Later studies recording the activity of single cells revealed more maps for different sensations such as temperature. Because of sensations transmitted from our skin to our brain, we can enjoy a gentle breeze and avoid the heat of a hot cup. We can distinguish different temperatures over a range from about minus 10 degrees to plus 60 degrees. Different sensory cells in the skin are responsible for transmitting this information to nerve endings. The mechanism is due to heat altering ion channels in these sensory cells, with different ones responding to a quite narrow range of temperatures. These heat-sensitive receptors, some for cold and others for heat, fire at a constant rate at the preferred skin temperature, but fire more slowly when the temperature changes. This is the mechanism by which cells tell us that we are hot or cold.

Pain is experienced due to receptors that respond to stimuli that can cause damage. Many respond to chemicals released by damaged cells, while others respond directly to nasty mechanical stimuli and extremes of temperature. To feel pain is necessary – it warns us of damage to our body. Pain is transmitted to the brain by electrical impulses in special nerves. These nerves have a special channel in the membrane for sodium near the site where pain is initiated; loss of or damage to this channel results in the inability of the nerves to send pain signals to the brain, which can be dangerous.

The subtlety of nerve cells can be illustrated by reference to those cells that transmit a painful stimulus from some part of the body to the brain. A young boy was found who felt no pain when a knife was plunged into his arm, and he died young from jumping off a roof while playing with friends. Examination of his relatives, who also could not feel pain, led to the discovery of a mutation in a particular gene. The mutation in the gene prevented the movement of sodium in pain nerves, which is necessary for the nerve to transmit its signal.

Seeing with our eyes begins in the retina, in two stages. The light that enters the eye is converted into an electrical signal by the retinal cells, and these signals are then passed along the optic nerve to the brain. The photoreceptor cells in the retina are of two kinds – rods and cones – which function respectively for night and day vision. A single particle of light can cause an electrical response in a rod but not in a cone, which requires more light. The outer regions of rods and cones are filled with light-absorbing pigments – photo pigments – attached to membranes. When light hits them they trigger changes in electrical charges across the membranes, prompting signals to nerves that connect with the optic nerve, which contains some million axons. These, as we will see, connect in a special pattern with a particular region of the brain involved with sight. Colour vision is based on three types of cones that respond to different wavelengths of light and thus to colour, and each has a different photo pigment. Absence or malfunction of these cones leads to colour blindness.

Our ability to hear and comprehend both a complex passage of music and the words of an ordinary conversation depends on what has been described as the almost miraculous feats of hair cells, the receptors for sound in the ear. Our hearing is due to sound waves causing the tiny bones in our middle ear to move and stimulate our fluid-filled inner ear, where thousands of hair cells convert the sound into nerve impulses. Each hair cell is most sensitive to stimulation at a particular frequency, and each is sensitive to only a limited range of frequencies. On average, neighbouring hair cells have a different characteristic frequency of about 0.2 per cent; by comparison, the adjacent strings on a piano differ by 6 per cent. Each hair cell directs its output to, on average, ten nerves. All the information must be sorted out in the brain. Loss of the hair cells or improper function is a major cause of the loss of hearing.

Smell involves a quite different set of cells. People who work on perfumes have claimed that they can distinguish between 5,000 different odours. Cells in the nose and mouth convey this information to the brain. In the nose are special nerve cells with a knob at one end from which some ten cilia protrude into the mucus coating the inner surface of our nose, while the other end connects with nerves that can relay information to the brain. The cilia have receptors that detect odours; there are about 1,000 different receptors for different odours. When an odour binds to the receptor it initiates a nerve impulse that goes to the brain. The nerve cells that detect smells are replaced at a rate of 1 per cent each day. They enable us to distinguish with ease some 10,000 different smells, and if you are a wine expert this could be as many as 50,000. Dogs have a tenfold greater ability. It thus comes as no surprise that 3 per cent of our genes work for these odour-detecting cells.

Taste is detected by cells in the tongue and palate, and wine gurus can distinguish more than a hundred different tastes. Four tastes are most easily distinguished: bitter, salty, sour and sweet. These nerve cells are most unusual as they are short-lived and are continually being replaced.

Almost all of our nerve cells are in our brain and spinal cord. As we saw, the spinal cord develops from the neural tube, and some of the cells in this tube begin to differentiate into motor nerves that will carry impulses to our body muscles, and sensory nerves that will bring information from all over our body. Those that will form nerves come from a division of the stem cells in the wall of the neural tube. Some of the divisions are asymmetric, so that one cell remains to divide again, while the other develops into a nerve and migrates away – the cells are acting like stem cells. Other divisions are symmetric: both cells could become nerves, or both could divide again. Repeated division and nerve migration during embryonic development lead to the formation of layers of nerves in our spinal cord.

The further development of the nerves determines where they will extend their axons, and with which cells they will connect. This is determined by their position. Motor nerve cells develop near the base of the tube. One key signal, our old friend Sonic hedgehog, is graded from a high concentration to a low one from the base of the tube to its top. The local concentration is thought to determine the positional value of the cells as in a French flag model, and at high concentrations of Sonic hedgehog motor nerves develop. There probably are conversations between the cells along the axis to ensure precision in specifying the cells' positions. The development of the cells is also determined by their position along the head-to-tail axis and is influenced by the Hox genes. Thus there are motor nerves that will send their axons to muscles on the upper side of the limb, while others will send their axons to muscles on the lower side.

Guidance of the extending axon is crucial in getting nerves to their targets during embryonic development. Early in nerve-cell development comes the formation of a growth cone at the tip of the developing axon. This cone guides the extending axon and pulls it along: like a migrating cell, it continually extends and retracts fine, finger-like extensions that attach at their tip to the surface they are moving over, their retraction pulling the growing axon forward. The extensions explore the immediate environment and determine where the axon will go. The cones are guided by molecules on the surface over which they are moving and to which they can adhere to varying degrees, and which can thus repel or attract them. Diffusing attractant and repellent molecules, sent to the cone from more distant sites, are also used to guide the cone.

This guidance mechanism can be illustrated by the way the visual system develops – how the optic nerve from the eye's retina makes the right connections to special regions of the brain. This has mainly been studied in frogs, but the same principles apply to us. The light-sensitive cells in the retina activate the nerves that form the optic nerve, and the nerves from the right eye go to the left side of the brain, while those from the left eye connect with cells on the right side. There is a point-to-point correspondence between a position on the retina, and the site on the brain. In a frog, if the optic nerve is cut and the eye inverted through 180 degrees so it is upside down, the axons manage to find their way back to their original sites of contact on the brain, but their worldview is now upside down: when presented with a fly, the frog moves its head in the opposite direction.

The mechanism for making the right connections is based on special molecules on the surface of the membrane of the growth cone interacting with other molecules on the membranes of cells with which the migrating cone makes contact. It is currently thought that both the optic nerve axons and the brain cells have positional identities, which allow them to match up. This enables the growth cones of the optic nerve, when it moves over the brain cells, to choose the right cell with which to establish a synapse. When migrating axons try to make contact at the wrong site, they are repulsed, and move on. The visual map in mammals is initially quite coarse-grained, but is later fine-tuned by activity of the nerves when the animal begins to see.

There is much cell death in the development of the nervous system. Some 20,000 motor nerves are formed in the region of the developing limb, but half of them commit suicide and die after they have developed. Their survival depends on their establishing contact with a muscle cell. Even if they do make contact, if more than one nerve contacts a muscle only one can survive. The nerves compete as to which will not commit suicide, and the survivor is the one with the strongest input into the muscle. The cleverness of the axons in finding their correct muscles is illustrated by turning upside down a piece of the spinal cord that sends nerves to the limb: even when the axons now enter the limb in quite the wrong position, they end up finding their correct muscle.

*

Turning now to the brain itself, it is remarkable yet again that in spite of all the different functions our brain carries out, the nerve cells in different regions are basically the same. What distinguishes one brain region from another is essentially the number of nerve cells and how they are connected – indeed, connection is all. Quite apart from the structural complexity of our brain, which comes from the special way the neurons are connected up during embryonic development, there are a thousand million synapses in a tiny piece of our brain the size of a grain of sand – and think how many grains of sand there are in our brains. It is most unlikely that we will ever understand the brain without new principles; even understanding quite simple neural networks can be difficult. When I argued, in a previous book, that science was unnatural and went against common sense, I did not imagine that our very essence lies within our billions of synapses. It is complexity at its most complex.

There are two overlapping stages in the development and maintenance of synaptic strength, the property that determines how our nerves interact with one another – and so how we think, learn, feel, and recall. The first stage is during embryonic development and is largely under genetic control. It is during the second stage that the fine-tuning takes place, largely through experience. Our memories are recorded in the nerve-cell networks in the brain, as is our ability to learn. All forms of memory and learning are due to the changes in the synaptic connections between nerve cells. Of particular importance is the ability of a large number of impulses to a nerve cell to increase the effect of signals at the synapse.

So how do we remember anything and how do we

learn? Nerve cells are still a bit secretive, but the answers lie in our synapses. A simple example of learning is habituation – that is, learning that a novel stimulus is harmless, so simple reflexes decrease when the stimulus is repeated. This mechanism is due to the diminished effectiveness of synaptic transmission from the sensory nerves to the motor nerves. Looking at the changes in synaptic connections in humans or other vertebrates is difficult, as there are so many cells involved. A marine slug, *Aplysia*, has thus been used instead to study its habituation to gill withdrawal when prodded. With repetition of the stimulus the synaptic potential in the responding nerves gets less, and so does gill withdrawal. This turns out to be due to a decrease in the number of transmitter vesicles released from the nerve at the synapses, and this change in synaptic vesicles is very likely the basis of short-term memory involving nerve-cell circuits. A related process is involved in how we and other animals can learn to respond more vigorously to harmful stimuli. Sensitisation occurs when a noxious stimulus is applied – this is due to enhancement of the synaptic transmission of the withdrawal reflex, as when, for example, we get an electric shock or a burn.

With more complex learning, other mechanisms are involved. An example of such learning is classical conditioning. Animals, ourselves included, can learn to associate one type of stimulus with another. This learning is dependent on the action being associated with a reward or a punishment: we learn to avoid hot things when they burn us, and to do well in our exams if our parents reward us with extra pocket money, and rats rapidly learn to press a lever if rewarded with food. This all involves synaptic changes. And long-term memory involves gene activity and protein synthesis that lead to new synaptic connections being made, and the pruning of some synaptic connections. There may also be changes in the synapses so that they transmit more, or less, neuro-transmitter.

We must hope that through the study of computer-based neural nets, networks that mimic connections between nerve cells, neuroscientists will begin to under-stand the computational principles that are involved in the behaviour of such networks in the brain. Neural nets can provide key insights into how our computer-like brains actually work. Neural nets are made up of inter-connected units, which behave like nerve cells. Their abilities, such as learning to recognise faces or handwriting, are remarkable. Banks in Finland have used them to check that the figure on the cheque that is being deposit-ed is the same as that written on the deposit slip.

Neural nets have their origin in the work of the math-ematical genius Alan Turing and other neuroscientists who were trying in the 1940s to model the brain as a computer. The nerve-like units in the net can be either on or off – they are like the components of computers which are based on being either an 0 or a 1. Each unit is con-nected to many other units, which it can either activate or inhibit. The key to the interactions between the units lies in the junctions between the units, which are weighted to determine the strength of the activation or inhibition. It is rather like saying 'Go' or 'Stop' to your neighbour either loudly or softly. Whether or not a unit then turns on is rather like voting: the unit looks at all the activation and inhibitions and their weightings, and decides who wins. The fundamental property of these junctions, which are analogous to the synapses between nerve cells, is that the weightings can change. The rule governing how the weightings change determines the behaviour of the net and enables it to learn.

A quite simple neural net can learn to recognise hand-written numbers from 0 to 9. For example, a hundred units are connected to 256 sensory units that read the image, and so there are over 25,000 junctions. The system is then presented with lots of 3s that are hand-written, and so vary a lot. This results in certain of the weightings junctions being strengthened while others are weakened. As a result, the system learns to use its hundred units as features that represent different pieces of the 3. The process is like using an identikit to represent a face; it uses a range of image fragments, like the mouth or an eyebrow. But the neural network learns all by itself what repertoire of fragments to use, and which ones to use for each particular 3.

Since the brain has so many nerve cells, it is surprising that much can still be learned about its function by examining the activity of single nerve cells one at a time, and even altering their activity. But such studies have helped us to understand how stimuli at our body's surface can be transmitted to the brain and analysed there. We need to add to this an understanding of how the complex circuits operate. Some memories are stored in the part of the brain called the hippocampus. A characteristic of its nerves is that when stimulated the increase in their response lasts for some time, which can result in increased protein synthesis. The hippocampus contains a map of the spatial environment in which we move. Our position is encoded by the firing of specific cells; only one cell will fire when we are in a particular position, and then another as we move.

There are those who argue that it is an illusion to think that some of our actions and those of our fellow humans are determined by our genes and our cells and not by free will. Genetics, it is asserted, is not destiny. I have a little sympathy with such beliefs, as it is all too easy to be mis-led as to what genes actually do for us, nor is it easy to accept that much of our destiny is due to our genetic inheritance. But just think about the differences between men and women. Only someone blinkered and unthink-ing could believe that all the differences are cultural. Shortly after we are conceived, the embryo from which we develop is indifferent with respect to sexual differen-tiation: female and male embryos look identical. Then, in males, a testis develops and secretes the hormone testos-terone and it is its effects on the embryo's cells, activating numerous genes, which make males different from females.

If you doubt that genes can determine criminality, look no further, for it is mainly males who commit violent crimes, not females. There is evidence of male superiori-ty in mathematically gifted children, and that in women, unlike men, language and spatial skills are located in both sides of the brain. From an early age the toys girls and boys choose to play with are different. It seems that female attachment to infants is innate, whereas with men it has to be learned. There are a number of structural dif-ferences in the female and male brain, which make sense from an evolutionary viewpoint as males and females have somewhat different roles to play in reproduction.

There is no equivalent in men of the female menstrual cycle and its attendant psychological and physical effects.

Why is there so much resistance to accepting that genes can play such an important role in our behaviour? It probably stems from what has come to be called the standard social science model of human behaviour, which has at its core the belief that most of our behaviour, and beliefs, are culturally determined, and that we are born with a brain that is in effect a blank sheet. In opposition to this, there is a model emerging in which biology is central to what the brain does and can do, but which does not deny the importance of experience and culture.

We are born with a much more sophisticated set of brain functions, all programmed by genes, than anyone previously thought. Very soon after birth we already have a concept of cause and effect in the physical world, and I have argued that this is what makes us different from all other animals and its evolution was necessary for human tool use. Our ability to learn a language involves a highly specific property of the brain, as do our mating desires. Many of us have an innate fear of snakes, but no child, no matter how often they are warned, fears an electric plug.

There is a need to understand how the brain is programmed by our genes to give us particular mental functions. Such knowledge could be helpful. If we understood the biological basis of criminality we might be able to do something quite new about it, and could treat people in the light of that knowledge. If there is a genetic basis for homosexuality it could reduce, for some at least, parental guilt, as well as psychoanalysts' fees.

A major unsolved problem with the brain is how it generates consciousness – that is, our awareness of ourselves and all the feelings and emotions and personal thoughts we have. How do those cells make us experience pleasure or sadness? The mystery remains.

10

How We Grow and Why We Age

how cells multiply, enlarge and decline

We, like most other animals, are born rather small and then grow. There is of course also growth during our embryonic development, but the embryo itself is very small compared to the adult into which it will develop. But our basic body pattern is laid down in the embryo when it is tiny – our limbs are only a few millimetres long when the basic pattern is already well established – and so are all our organs. The basic patterning mechanisms mainly involve interactions between the cells over less than a distance of 30 cells.

If one looks at a newborn baby, it is clear that different parts will grow to different extents. The head of the baby when it is born is much larger compared to the rest of the body than when the child grows up. How much each bit will grow is partly programmed in the cells in the embryo at an early stage.

We start looking like a little embryonic animal when we are in the womb and are about 1.5 cm long, and at birth have increased to about 50 cm; then we grow to around 180 cm, varying between males and females, with growth ceasing after puberty. Insulin-like growth factors play an important role in this growth, as well as growth hormone, which is made in the pituitary gland of the brain.

There are three ways by which cells cause regions to grow bigger. The main one is, of course, cell multiplication. But cell enlargement is also important. Just consider the long extensions from nerves: once a nerve cell is born, it will never divide again, but it can grow very extensively in size. Similarly, muscle cells never divide, but they can grow and also fuse with special stem-like cells associated with them to make large muscles. The third way is for the cells to add material to the space around them, as in the case of cartilage and bone. But again, even when organs grow there can also be considerable cell death.

Both internal controls and external signals are important in determining growth and ultimate size. The role of internal controls is illustrated by the developing spleen. If several embryonic spleens are placed in an early mouse embryo, each spleen will be much smaller than normal, and the total size of the little spleens will be that of a single normal spleen. This means that there are circulating inhibitory molecules which determine how large the spleen grows. Such factors could be made and sent out by the spleen and will inhibit growth as their concentration rises, so that the size of the spleen will depend on the overall size of the animal. This is also true of the liver, and even if as much as two thirds of the adult liver is removed it will grow back to its normal size because the inhibitory factors it makes have been reduced. The ability of the liver to regenerate is nicely illustrated by the Greek myth of Prometheus. Having angered Zeus, he was chained naked to the side of a mountain, where a vulture

tore out and ate his liver, which regenerated. One won-
ders if the Greeks really knew about the liver's remarkable ability to restore itself.

By contrast, the cells in some organs, like our limbs,
control the size of the organ and are not at all interested
in whether there are other similar organs growing nearby. This is also true of our single thymus gland, which is
located in our chest and is part of the immune system. If
several thymus glands from embryos are transplanted
into a developing mouse embryo each will grow to its
normal size, as will the normal thymus gland of the host
mouse. A similar growth mechanism is found in the pancreas, which contains the cells that secrete insulin, essential for the entry of glucose into cells. If the number of
cells in the developing pancreas is reduced, then only a
small pancreas will develop. It seems that the cells of the
developing pancreas are programmed to grow through a
fixed number of divisions and there is no regulation of
size in relation to the rest of the body, or the possibility
of regeneration. This has important implications for diabetes. Why there should be these different control systems for organ growth is not clear.

Growth of the embryo in the womb has important
implications for our health. This was established in a
remarkable set of studies by David Barker, who discovered that children born with low birth weight had an
increased chance of heart disease in adult life. Animal
studies have shown that the mechanism involves changes
in gene expression in the embryo in response to its environment. The key factors are those from the mother's
nutrition and lifestyle, and whether she is fat or thin. The
embryo can predict that in its future world there will be

too little food – and this is reflected in the development of its circulation.

It can occur that too much food is brought to the cells by our blood transport system – which simply transports what is there, and much of this comes from what, and how much, we eat. A little more than half of the adults in the United States are overweight or obese; though there are genes associated with obesity, as shown by examining the DNA of obese individuals, this is mainly due to overeating. Obesity is due to an excess of fat contained in fat cells. It is associated with numerous diseases, including Type 2 diabetes and heart disease. Our body fat is made of some 40 billion fat cells, with most of the bulk stashed under the skin. By contrast with other body cells, in fat cells the fat occupies some 95 per cent of the cell volume and other structures, like the nucleus, are shoved against the cell membrane. White fat cells contain a large fat droplet and the fat stored is in a semi-liquid state, while brown fat cells have fat droplets scattered throughout. Brown fat, also known as 'baby fat', is used to generate heat.

The amount of fat someone has is a reflection of both the number and the size of the fat cells. We are born with a number of fat cells, with women generally inheriting more than men. The number of fat cells increases through late childhood and early puberty, and after that they remain quite constant in number. Fat cell numbers increase more rapidly in obese children than in lean children and thus obesity in adulthood is linked to childhood obesity. People get 'fat' by filling up all those greedy fat cells – obesity thus develops when a person has too many fat cells and they increase in size. People with extra fat cells may be able to shrink these cells by dieting and exercise and get great results, but their fat cells don't just disappear. Once fat cells develop in the body, they remain there for life, and they seldom die, though there is some turnover: each month about one in every hundred of our fat cells dies and is replaced. Once the fat is burned from the cell they become much reduced in size but these greedy cells are just waiting to be fed and to get fat again, so it is not easy to maintain weight loss.

Fat cells are involved in monitoring energy reserves, and are the hub of a complex communication system that regulates many functions, continuously telling the brain how much energy the body has left, signalling muscles when they can burn fat, instructing the liver and other organs when to replenish fat stores, and controlling the flow of energy in and out of cells. Fat is dynamic, exchanging chemical signals with the brain, bones, gonads and immune system, and with every energy manager on the body's long alimentary track. Key among the chemical messengers is the protein hormone leptin, an essential player in obesity. Leptin circulates in the blood and is made by fat cells, and the bigger the fat cells the more leptin they make. Leptin tells the brain when we have eaten enough and lessens hunger, but it has to compete with other brain functions that lead to greed and pleasure.

An example of a system being programmed early on to grow at a particular rate comes from the arm. When the arm is first formed in the developing embryo, the wrist elements are the same size as all of the more proximal elements – the humerus and radius and ulna – but they will grow much less, and so end up much smaller. The obvious big difference in the size of the bones in our limbs reflects differences in growth rate, which is programmed early in development.

But how do our limbs actually grow? The future bones in the limb are initially laid down, as we have seen, in the embryo as cartilaginous rods. This cartilage then begins to be replaced by bone in the centre by special cells, and this spreads towards the ends. Near the ends bone ceases to be made and a growth plate is formed which will become the region of bone growth. In bones like the humerus there are growth plates at each end. Thin growth-plate cells are arranged in columns, and at the top end of the plate are stem cells; the next set of cells are multiplying ones, and further along they stop dividing and become cartilage cells. The cartilage cells increase in size as we move further down the growth plate, and at the end of the plate they die and are replaced by bone. Thus the growth plate lengthens the bone by the increase in size of the cartilage cells and their replacement by bone at one end, and the number of cells in the column remains the same due to stem cells adding to it at the top of the plate. Growth hormone stimulates the stem cells to provide cartilage cells and other hormones stimulate cell division in the region where the cells are multiplying. So our limb growth is due to cell multiplication and cell enlargement, which is then made into bone. The growth plates are different in different bones, and so their rates of growth differ.

Growth plates continue to increase the length of our bones until puberty, and their closing down may be due to hormonal changes. Our hormones play a major role in how long growth continues. Another possibility is that the growth plate is programmed to grow for only a specified time, or for a certain number of cell divisions. If the number of divisions in a plate is reduced for a time, there is then catch-up growth, with more cell divisions making up the loss.

Our two limbs grow from less than a centimetre in the embryo to a hundred centimetres in the adult and they have very similar lengths, matching with an accuracy of about 0.2 per cent. A striking feature is that there is no known connection between the limbs during those fifteen or more years of growth, nothing that tells one limb about the growth of the other. The growth of our limbs is usually very reliable, and we do not know how this is achieved. A typical limb growth plate is about 2 cm in diameter and less than 1 cm deep; there are some 10 million cells in the plate arranged in 300,000 columns, each of which is about 40 cells long. Typically, each day one cell is added and one replaced by bone in each of the many columns. The large number of cells and the mechanical constraints may make the growth rate for the plates similar for the two limbs.

The muscles in our limbs must also grow, and muscle cells, once formed, cannot divide again. But the muscle cells grow a great deal by just getting bigger, as well as by their fusion with special muscle satellite cells that bring in additional nuclei. These satellite cells are a sort of muscle stem cell. How much the muscles grow largely depends on the growth of the bones to which they are attached. As the bones grow, the muscles are put under tension and are stretched, which causes them to grow. The growth of muscles and bones is therefore coordinated, and the muscles do not have to know anything other than how much they are being pulled. Muscles continue to respond to tension, and use and pumping iron with heavy weights will increase their size.

Our growth after birth is most often dependent on milk from our mother's breasts. Milk production is switched on when the baby is born and then turned off when the baby is weaned. The production of milk by cells in the breast is due to hormones that circulate at birth. The hormones present during pregnancy cause the cells lining the ducts in the breast to multiply so that there are some twenty times more of them. At their tip are the cells that will secrete milk. The baby sucking at the breast stimulates cells in the brain to release a hormone which causes the cells in the breast to squirt out milk. Once breastfeeding stops, the cells that secrete the milk die by apoptosis and their remains are cleared away by those hard working macrophages that get rid of any unwanted material.

We are born with a brain containing around 100 billion nerve cells, and there is little further cell division after birth. The weight of our brain when newborn is approximately 300 grams, about 10 per cent of body weight – in contrast to the adult brain, which, weighing approximately 1,400 grams, is just 2 per cent of body weight. Brain size increases with age and acquires its adult maximum size between six and fourteen years of age. The increase in size is due not to multiplication of nerve cells, which do not divide after they have become fully specialised, but to supporting cells, and to the growth of the nerve cells as they branch and make new connections. New synapses are formed at a rapid rate during the early months of life, usually achieving maximum density between six and twelve months after birth. There is a decrease after this, due to disuse or natural attrition – when formed, the infant's brain retains only those synapses that it frequently uses. Early sensory experiences are thus vital for the formation and retention of synapses.

Most of us would prefer to avoid dying, but ageing and its attendant pains may not be preferable. Recall the Greek myth about Tithonus, who was a lover of the goddess Eos. Eos asked Zeus to grant Tithonus immortality. He granted her wish, but Eos neglected to ask that youthfulness should also be given. Tithonus aged beyond recognition and begged to die.

If cells are so clever, why do they and we age? When we have passed sixty only about 25 per cent of the variation in life span is genetic. The blame must fall on evolution, which is interested only in reproduction and not in health once we have reproduced. Any active mechanical system ages because of wear and tear, and this is true too of the cells in biological systems. Once past our sixties and into our seventies and beyond, the effects of ageing become all too obvious. Our memory is not what it was, we run more slowly, our knees hurt – the list seems endless. And there is the puzzle as to why animals have such different life spans: mice can live only some three and a half years, while the finback whale can go on to eighty. But there is little evidence that ageing makes a significant contribution to mortality in the wild, as other factors such as predators and illness will cause death earlier. More than 90 per cent of wild mice die in their first year.

Ageing is not part of our developmental programme and there are no genes that promote ageing. On the contrary, evolution has sensibly found cell activities that prevent ageing – but these are active only until reproduction is no longer a significant activity. The effect of evolution can be seen by comparing two-year-old mice with baby elephants at the same age. The mice are already old. Evolution has selected mechanisms to prevent the elephant ageing before it has offspring, and for some elephants old age is simply worn-out tusks. From the point of view of evolution, the prevention of ageing is only necessary until the animals have reproduced and cared for their young; nature has therefore provided repair measures to delay the process until that is done. This is the 'disposable soma' or disposable body theory – we and other animals are disposable once reproduction and the rearing of children have been completed.

But if ageing is not programmed by our genes, then why does it happen? For the same reasons, basically, as in our cars and washing machines: wear and tear. Ageing results from an accumulation of unrepaired cellular and molecular damage and the limitations in cell maintenance and repair functions, particularly in our DNA and proteins. The maintenance of the integrity of DNA is a challenge to every cell, for damage leads to the absence of key proteins, to the synthesis of proteins in the wrong cells at the wrong time, and also to proteins with bad properties. Such damage accumulates throughout life from the time when body cells and tissue first begin to form. Other damage occurs in mitochondria and membranes and from the misfolding of proteins. How long we and other animals live is determined primarily by mechanisms that have evolved to regulate the levels of body maintenance and repair functions. Our germ cells, which give rise to eggs and sperm, require elevated levels of maintenance and repair, as the ageing of germ cells would be disastrous for reproduction, but this is not so for body cells.

There is no question that our cells age. If you take some simple connective-tissue cells from your body when you are young and put them in culture they will happily multiply for a while, and then stop. The number of multiplications will be around fifty. If the same experiment is done forty years later, your cells will divide far fewer times, around thirty. The reduced number of divisions reflect an ageing process. Yet there are some cells in our bodies that do not show this ageing, and seem happy to continue to multiply for long periods. This seems to be true of stem cells. By contrast, cells from an individual who has Werner syndrome, which results in premature ageing, divide fewer times in culture than normal. People with the genetic defect that results in Werner syndrome age rapidly after puberty and have a short stature; by the age of forty they appear severely aged, and they usually die in their late forties of cancer or heart disease. Just one gene is involved, but it is required for DNA replication and repair and this may be the cause of the premature ageing.

One explanation for the decline in cellular division capacity with age appears to be linked to the fact that the telomeres (from the Greek word for 'end part'), which protect the ends of chromosomes, get progressively shorter as cells divide. This is due to the absence of the enzyme telomerase, which makes the telomere grow back to its normal length after division. This enzyme is normally expressed only in germ cells in the testis and ovary, and in certain adult stem cells, as these cells have to be prevented from ageing. It may be that the telomeres can count how many divisions the cell has gone through, as they get a little shorter at each division. This could function to protect the cell against runaway cell divisions as happens in cancer, and ageing could be the price we have to pay for this protection. In diseases that result in premature ageing there is accelerated telomere shortening, and this may be partly responsible for the condition.

A number of genetically based diseases show accelerated ageing, like Werner syndrome. A very rare one is progeria, which is due to a protein defect, and those who have it die as young as thirteen and already look old. The affected protein is one associated with the membrane of the cell nucleus, which is involved in DNA and RNA synthesis.

Chromosomes may be the structure whose integrity cells have the most difficulty in maintaining over their lifetime. The DNA in every chromosome experiences thousands of chemical modifications every day. Several of the most important mechanisms causing ageing involve damage to DNA, and there is a general relationship between longevity and DNA repair. Our cells try to avoid this damage by wrapping the DNA around a set of proteins, and the more dense the packing, the more the DNA is protected – but this can make it more difficult for a gene to be transcribed and then translated into a protein.

Cells tend to respond to dangerous DNA damage by committing suicide – apoptosis – and this provides a way of preventing the damaged cell becoming cancerous. This occurs much more often in aged tissues in which the background accumulation of damage is greater, and the resulting loss of cells may itself accelerate senescent decline. Long-lived organisms invest in better DNA maintenance. The benefit of this is seen both in slower ageing and delayed incidence of cancer, since genome instability contributes to both these processes. Humans are less likely to get cancer than mice as they have invested more in DNA repair. But while long-lived organisms make greater investments in cellular maintenance and repair than short-lived organisms, with age the repair mechanisms fade.

Nature and evolution had a fine sense of irony when they made our lives so dependent on oxygen – which is essential for energy production, but is also a major cause of ageing and our eventual death. A possible cause of damage to DNA and other molecules is reactive oxygen species, which are small modified oxygen molecules. Oxygen is required by the mitochondria in cells to produce energy from the molecules derived from food; production of ATP by the mitochondria also results in the production of these reactive oxygen molecules, which are potent agents that cause damage to many targets in the cell, including DNA. They can also damage the mitochondria, leading to less energy production, itself a characteristic of ageing.

Protein turnover is essential to preserve cell function by removing proteins that are damaged or redundant. Age-related impairment of protein turnover is indicated by the accumulation over time of damaged proteins, and there is evidence that an accumulation of altered proteins contributes to a range of age-related disorders, such as Alzheimer's and Parkinson's disease.

Evidence from animals shows that limiting food intake – in other words, eating less – can significantly extend the lifespan of a variety of animals. When rats are kept in the laboratory under pleasant conditions but with an intake of food such that after weaning they get 50 per cent less than their well-fed neighbours, they live about 40 per cent longer. The oldest rat with high food intake is around a thousand days old, but there are those on the restricted intake who get to 1,500 days. In female rats the age at which the ability to reproduce is lost is extended from 18 months to 30 months. Vitamins and minerals must be included in the diet but it does not matter if the reduced calories come from carbohydrates, proteins or fat. Low intake of calories suppresses most of the diseases so common in older animals like cancer, high blood pressure and deterioration of the brain. If the restricted feeding regime is returned to full feeding, the ageing process then seems to be actually accelerated.

There is some evidence that we humans could also delay ageing by reducing our calorie intake. On the Japanese island of Okinawa there are significantly more centenarians than on any other Japanese island. The death rates from stroke, heart disease and cancer are only about two thirds of those for Japan as a whole, and the death rate for sixty-year-olds is half the national average. It is unlikely to be just a coincidence that the average adult food intake is, for cultural reasons, 20 per cent less than the Japanese average, and that schoolchildren on Okinawa eat less than two thirds of that recommended.

That such increased ageing may also be related to insulin comes from studies on worms and flies. A dramatic example of an increase in lifespan comes from the nematode worm, which has about half our number of genes and normally only lives for about 25 days. If the worms are placed under conditions where there is a limited food supply and many other worms are present, then instead of developing into adult worms through a series of larval stages, they develop into an alternative larval form known as a dauer larva. These dauer larvae neither feed nor reproduce, but, if conditions improve, moult into adulthood and can then reproduce. The dauer larvae, with their very dull lives, can live for up to 60 days, more than twice as long as normal worms. This is due to interference with the insulin pathway: if a gene in the worm that codes for a receptor for insulin like protein is knocked out, the worms live twice as long. The mechanism is not clear, but involves many other proteins. Reduced signalling by chemicals similar to our insulin also extends the life span of the fly *Drosophila*. It has recently been shown that in mice, less insulin receptor signalling throughout the body, or just in the brain, extends lifespan up to 18 per cent. The effects of eating less may operate via the insulin effect. Fasting reduces insulin secretion, but one must be cautious in trying too hard to reduce insulin secretion as this can lead to diabetes.

A few primitive multicultural organisms, including hydra, a primitive simple animal in the form of a tube with tentacles, exhibit very slow or negligible ageing. Individual animals observed over a period of four years showed no age-related deterioration, either in terms of survival or reproduction rates. The reason is not clear, but may be related to the fact that hydra can reproduce by forming buds which will develop into mature hydra without sexual involvement, and are also capable of undergoing complete regeneration from almost any part of their body. Most of their body cells can contribute to regeneration, so if some age they may die or be lost at budding.

Lucky hydra, as humans can suffer severe mental problems in old age. Alzheimer's disease is amongst the worst. It was in November 1906 in Tubingen, Germany, that Alois Alzheimer presented to the world the nature of the disease that was later named after him. He had examined the brain of a woman who had suffered from progressive memory loss and increased confusion, and he used a new method for looking at sections of the brain under the microscope. He found abnormal deposits in the brain, which were much later identified with the electron microscope as tangles of filaments and plaques. A small protein, amyloid beta, gives rise to the plaques and another, the protein tau, to the tangles. The disease arises from damage to and the death of nerves in specific regions of our brain. Just why this happens is not clear, but there is a genetic component which leads to degeneration of the nerve cells that can start some twenty to thirty years before there are any symptoms. A nasty problem, as early signs are loss of some commonplace memories like names, as many of us know all too well. Around half of those over eighty years old show some symptoms of Alzheimer's or other forms of mental dementia.

Cells can respond to new situations by increasing their growth in ways that do not lead to cancer, but have other effects. All too good an example is the prostate in older men: the prostate protrudes into the bladder and causes increased urination. This is due to increased proliferation of the epithelial cells and fibroblasts, and an increase in size of the smooth muscle. Why this should occur with age is just not known, and it can develop into a tumour. But not all abnormal increases in growth are cancerous.

A peculiar and unsolved problem is whether anyone actually dies from old age. There is almost always a good explanation for anyone who dies when very old in terms of the abnormal behaviour of their cells.

A question not often asked is: 'When does death begin?' Perhaps at birth. Some two thousand years ago Marcus Aurelius wrote: 'Mark how fleeting and paltry is the estate of man – yesterday in embryo, tomorrow a mummy or ashes.'

11

How We Survive

how cells defend against bacteria and viruses

When we get ill, it is in fact our cells that become abnormal or sickly. Our bodies have evolved mechanisms to defend against invaders like bacteria and viruses who are searching to find a good place to reproduce, and against the physical damage to our cells that comes from cuts and burns. Special cells move into the affected area and try to repair the damage. The cells of the immune system, which can recognise foreign bodies as distinct from those normally found in the body, play a key role in dealing with these problems.

Bacteria and viruses are the main biological enemies that our cells must defend themselves against. It has been estimated that infectious diseases cause about one third of all human deaths; the AIDS virus alone has caused some 20 million deaths worldwide. These infectious agents usually exploit the environment within the cells they infect, and do so as cleverly as the cells they damage. They love our warm cells, which contain all the nutrients and elements for their reproduction, and many have evolved mechanisms to try to avoid our cells' attempts to eliminate them. Some bacteria can only grow when they are inside our cells, while others can do so without entering them. They also cause damage locally, which helps them to spread to new sites.

Bacteria are the simplest forms of life as they contain no internal structures like a nucleus to contain their DNA, nor mitochondria to produce energy. They are typically spherical or rod-shaped, and often have a moderately rigid cell wall. They can reproduce in just 20 minutes, and so in eleven hours one bacterium can give rise to five billion new cells. We have ten times as many bacterial cells in our body as we have normal body cells – a fact so surprising that you may need to read that again. In spite of their enormous numbers these bacterial cells are so small that they are in no way visible to us. Every time we speak we release thousands of bacteria into the air.

A major reason why we put up with this invasion, and allow these bacterial cells to make us their home, is that they prevent the entry of many really nasty bugs, as they have already occupied those desirable niches between our cells. In our gut, they are essential for digestion. How we distinguish them from nasty invaders is an important question for our immune system. Babies in the womb have no bacteria to deal with, but a short time after they are born their gut becomes home to one of the densest population of bacteria on this earth. Entry is possible, for example, via the nose, anus, mouth and any damaged region. There are also many bacteria on barriers to entry like the skin. Internal surfaces have special mechanisms to keep them out. The cells lining our airways have, as mentioned earlier, a layer of mucus which is swept out by cilia.

While there are good features associated with this population of bacterial invaders in our gut, as they are essential for the digestion of our food, they are also responsible for several illnesses such as inflammatory bowel disease and colon cancer. Stomach ulcers are caused by a particular bacterium, and this was a finding that went against the accepted view. The brave doctor who discovered this proved his point by drinking a pure culture of the bacterium, from which he developed a stomach ulcer. Only some bacteria are the cause of damage, and this is often because they secrete substances like toxins that cause the disease and allows them to spread and increase in number. Cholera is caused by just such a toxin, which is responsible for the severe diarrhoea characteristic of the disease. The bacteria colonise the small bowel and produce a toxin that causes loss of fluid from the gut, with serious consequences. Because they are sensitive to the acid in the bowel, an enormous number of bacteria must be swallowed to bring about the disease.

Before we look at the immune defence system and how it deals with invaders, we will look at the more simple defence mechanisms involved in wound healing. This introduces us to inflammation, which is a feature of what is known as the innate immune system. We will then consider the adaptive immune system, which makes antibodies.

Cells are good at healing wounds like a cut in the skin, provided the wound is not too large. The damaged blood vessels release blood which clots due to the action of platelets, which aggregate and bind to collagen protein fibres to form a platelet plug which will later give rise to a scab. The platelets also release a variety of factors that help to stop bleeding at the wound. Neutrophils, white cells from the blood vessels, migrate into the damaged area, and the epidermal skin cells multiply and grow beneath the scab. Macrophages move in and eat up dead cells whose contents could damage neighbours and interfere with healing; this process, called phagocytosis, involves responding to external signals and membrane remodelling. The cell sends out extensions that surround the object to be phagocytosed, and then totally engulf it.

After five days, cells that have entered the wound area have laid down bundles of collagen and there is a scar. Skin cells begin from an early stage to advance across the wound and meet their friends from the other side. But if the loss of the skin is more than 2 inches in diameter, and quite deep, it will not heal well and will need a graft. Skin can be taken from other parts of the body, and there are now techniques for growing skin cells in dishes and then grafting them on to the wound.

A different repair process takes place when a bone is fractured. The healing process is mainly determined by the connective-tissue membrane covering the bone. It supplies most of the cells which develop into cartilage and is essential to the healing of bone. The bone marrow (when present), small blood vessels, and fibroblasts are secondary sources of precursor cells. Soon after fracture, the blood vessels constrict, stopping any further bleeding with a blood clot, and then all of the cells within the blood clot degenerate and die. Days after fracture, the cells form some cartilage which is replaced by bone by special cells – osteoblasts – which have their origin in the bone marrow together with blood cells.

Inflammation is an early response of the innate

immune system, and is the mechanism by which the body provides a startled defensive reaction to almost any trauma such as the entry of bacteria or viruses. It results from many infectious diseases, from physical damage or from problems caused by excessive heat. Several aspects of wound healing are involved in inflammation, and the vascular system and white cells are also intimately involved. Many injuries lead to inflammation – as was first described in Egypt around 300 BC – and also the following illnesses: asthma, gout, multiple sclerosis, atherosclerosis, and probably Alzheimer's disease. It causes local cells to call for help from other cells, including those of the immune system and others.

When we are invaded by nasty bacteria and viruses, inflammation helps to contain the offensive agent and to eventually eliminate it, but the inflammation itself can often cause damage to the local tissue, as in tuberculosis. Damage to our tissues by wounds or invasion by bacteria or viruses releases three types of messages for producing inflammation. The first response to local pain is that the local nerve cells secrete signal molecules, following which any broken cells release proteins that call for help, and then, if bacteria are sensed, this results in a further plea for help. The redness associated with inflammation is brought about by a change in the microcirculation leading to a massive outpouring of cells and fluid, which can cause a swelling. The redness associated with an inflammation is due to the increased flow of blood. Fluid loss is easily recognised when we have a cold in the nose caused by a viral infection which leads to a local inflammation.

Among the cells involved in the call for help when the innate immune system is activated are mast cells, and these also contribute to inflammation. The mast cells release substances that increase the blood flow and so cause redness. Mast cells are scattered throughout the skin and provide an early response to infection, secreting key signalling molecules. Mast cells have been called chemical factories: they produce some 10,000 different molecules that can make blood vessels leaky, and activate immune cells. The outpouring of their contents can make our skin itch and our noses stream, and even provoke suffocating asthma attacks. But – and this was only recently realised – they are essential for fighting infection as they summon lymphocytes to the site of infection, and help to activate the immune system. On the bad side, they can help to initiate rheumatoid arthritis and are involved in the destruction of the insulating cells (Schwann cells) of nerves, which causes multiple sclerosis. In cancer, they can help to attract blood vessels to the tumour. All this local activity by the innate immune system can cause damage to local normal cells, so there need to be mechanisms to stop tissue damage when the bacteria are no longer a problem, and to promote healing.

Appendicitis is an example of acute inflammation. There is pain and often nausea and fever. What initiates the condition is not clear, but usually some obstruction causes a build-up of pressure which affects the blood supply and damages the wall of the appendix, which is then invaded by bacteria. Inflammation is the result.

An early response that leads to inflammation is the increased blood flow at the site of injury and a loss of fluid from the fine vessels, together with an accumulation of red cells. Changes in the vessel walls enable leukocytes in the blood to move out, squeezing between the cells of the vessel, and then starting their work of destroying infectious agents and eating up pieces of dead cells. Leukocytes are white blood cells of the immune system; they include cells like macrophages, whose function is to devour invaders and clean up any debris. They will migrate to the site of damage by following a chemical signal in the direction of increasing concentration of a diffusing agent. This agent is often a product of the invading bacteria. When they get near the site they can even move across cell sheets to get to the bacteria, which they then engulf and kill. They have a mechanism for recognising such foreign bodies, and also release enzymes that can attack infectious agents. But they can also damage normal healthy tissue in the immediate environment. It is not easy to kill invaders without damaging some of your friends, and tuberculosis illustrates this.

Tuberculosis (TB) is caused by a bacterium, as was discovered by Robert Koch in 1878. Over one third of the world's population have been exposed to the TB bacterium, and new infections, it is claimed, occur at a rate of one per second worldwide. But the bacterium does not cause its serious effects by producing any nasty toxins – it is our body's response that causes the disease. When the bacteria enter the lung from the air they usually cause a local lesion about 1 centimetre in diameter where they multiply. The macrophages naturally respond to this local intrusion and eat the bacteria. Then T lymphocytes (T cells), which will be discussed below, arrive and bind to the macrophages, causing the cells to kill the bacteria inside them. All this causes a local inflammation in which many normal lung cells die. The bacteria can also spread to other parts of the body. Another nasty disease is pneumonia, which can be caused by either bacterial or viral infection of the lungs. The bacteria or viruses invade the lung and cause local inflammation, which fills the lung's air spaces with fluid and so prevents oxygen uptake. These respiratory infections are spread by droplets and kill many children worldwide.

There are many other diseases caused by bacteria, some of them very serious. Plague, also known as Black Death, has caused more deaths in history than most other diseases. The bacteria that cause plague are almost always inoculated into the skin by a flea-bite, and then migrate to the lymph nodes. The fleas live on rats, so people acquired plague only if they were in contact with recently dead rats. Only rarely is the infection passed from person to person. Admittedly, plague is now so rare that almost all modern cases are due to isolated instances of transmission from animal reservoirs, but bubonic plague was probably common in the historical great plagues – it acts very fast and explains the stories of people just dropping dead suddenly, without the usual signs. The bacteria that cause diphtheria secrete a toxin that blocks protein synthesis; this results in the progressive deterioration of myelin sheaths of the nerve cells in the central and peripheral nervous system, leading to degenerating motor control and loss of sensation. The bacterium is spread by airborne droplets and can survive drying. Tetanus bacteria are motile rod-shaped organisms that are common in the soil. If they infect a deep wound they secrete a toxin that causes muscles to contract and go into spasm. The toxin binds to the synapses of inhibitory nerves and so motor nerves, freed from inhibition, initiate an abnormal contraction of muscles. Leprosy bacteria can affect many organs: skin, eyes, bone and testes. Syphilis is sexually transmitted and while it starts as a small sore, it can later spread and damage many organs. Tooth decay, a very common effect of bacteria, is caused by certain types of acid-producing bacteria which cause damage when sugars are present. The increase in acidity affects teeth because it reduces the tooth's special mineral content.

The discovery of penicillin, which can destroy many bacteria, was made by Alexander Fleming in 1928. Fleming, at his laboratory in St Mary's Hospital in London, noticed a halo of inhibition of bacterial growth around a contaminant blue-green mould on a bacterial plate culture. He concluded that the mould was releasing a substance that was inhibiting bacterial growth and killing the bacteria. When he grew a pure culture of the mould he discovered that it was the fungus *Penicillium*. Fleming coined the term 'penicillin' to describe the filtrate of a culture of the mould. Even in those early stages, penicillin was found to be most effective only against certain bacteria, and ineffective against other fungi. He expressed initial optimism that penicillin would be a useful disinfectant, though after further experiments Fleming was convinced that penicillin could not last long enough in the human body to kill nasty bacteria. In 1939, however, the Australian scientist Howard Florey and a team of researchers at the University of Oxford made significant progress in showing that when penicillin is injected into an animal it is very effective in killing bacteria. Their attempts to treat humans failed due to insufficient volumes of penicillin, but they proved its harmlessness and its effectiveness in mice. In 1942 they became the first team in the world to successfully treat a patient using penicillin. Penicillin was being mass-produced by 1944. During World War Two, penicillin made a major difference to the number of deaths and amputations caused by infected wounds amongst Allied forces, saving an estimated 12 to15 per cent of lives. Penicillin works by interacting with certain proteins in the bacterial cell wall and so preventing bacterial cell growth leading to death of the cell. But it has no effect on viruses.

Viruses are potent infectious agents. Viruses contain some of the key molecules present in cells, but they have no ability to reproduce on their own – they are not alive. They can only reproduce by getting into cells and using the cells' machinery. They have been called chemical zombies, but they can cause serious trouble if they invade our cells. They are very small and can pass through ultra-fine filters. They contain relatively few genes – varying in number from three to a few hundred – enclosed in a protein coat. In addition to the nucleic acid for the genes, which may be either DNA or RNA, they also contain three classes of proteins: proteins for their replication, proteins for their structure, and proteins that will affect and usually damage the cells they infect.

When viruses infect cells, the cells desperately try to prevent them replicating. Viral replication makes use of a peculiar molecule, a double-stranded RNA. One way the cells attempt to eliminate this is by the production of the protein interferon, which stimulates the activity of a gene that breaks down RNA and thus prevents viral replication. But in destroying the viral RNA, there is the danger the cells' own RNA will also be destroyed, and in some cases the cell destroys itself in order to prevent the virus replicating. Interferon also enhances the activity of certain killer cells of the adaptive immune system, which kill infected cells. But many viruses manage to get past these defences.

The first step for a viral infection is for the virus to bind to the cell surface. Entry into the cell by some viruses is by fusing with the cell membrane and then going inside – this is the case with the AIDS virus. Entry by other ways is more complex, and seems to be by the virus making a pore in the membrane so that its genetic material can enter the cell. When they enter the cell, viruses use the cell's machinery to read their genes and synthesise their proteins. Many viruses possess proteins that result in the cell favouring the synthesis of the viral proteins rather than that of the host cell. The protein and the nucleic acid of the virus reassemble in the cell to form many viruses.

The reproduction of the virus in the infected cell can lead to the cell breaking open and releasing many more viruses to infect nearby cells. This occurs when we get a cold (there are some hundred cold viruses): the viruses cause a local inflammatory response, and thus the dilation and leakage of blood vessels and mucus gland secretion which is typical of colds with blocked noses. Inflammatory mediators also activate sneeze and cough reflexes and stimulate pain nerves. The cold virus likes to infect the cells lining our nose and windpipe, and the common symptoms of flu such as fever, headaches and fatigue derive from the huge amounts of inflammatory substances produced from the infected cells, which give rise to the symptoms but cause no damage to the tissues.

In contrast to the virus that causes the common cold, the influenza virus can cause tissue damage, so symptoms are not entirely due to the inflammatory response. In more serious cases influenza can cause pneumonia, which can be fatal, particularly in young children and the elderly. Before 1900 flu in humans was a mild illness, but a different strain infected chickens and ducks, which together with a human strain infected pigs, so that a new very virulent strain evolved. It was this strain that caused the flu epidemic in 1918 which killed some 40 million people worldwide – more than had been killed in the First World War. The high virulence of this flu virus was due to its causing the immune system to over-react: the signals summoning immune cells to the site of infection became so strong that the site was over-whelmed with immune cells, and this blocked airways and damaged tissues.

Some viruses have genes encoded in RNA rather than DNA. When they enter the cell their RNA is translated into DNA, and that DNA is used to code for proteins to make new viruses. The AIDS virus has its genes encoded in RNA and the DNA that is made can hide in the host's DNA in the chromosome; this makes it more difficult to treat with anti-viral drugs. Measles is caused by a highly contagious airborne RNA virus that spreads primarily via the respiratory system. The virus can be passed from person to person via droplets in the air containing virus particles, such as those produced by a coughing patient. Once transmission occurs, the virus infects the skin cells of its new host, and may also replicate in the urinary tract, lymphatic system, blood vessels and central nervous system. Rabies is due to another RNA virus; it may replicate in muscle cells or nerve endings, and can travel down nerves to the central nervous system and cause flu-like symptoms which then lead to partial paralysis, dementia, and death.

Antiviral drugs can act by binding to the receptor on the cell membrane to which the virus itself must bind in order to enter the cell; or by binding to that part of the viral coat that binds to the membrane, so that the virus cannot attach to the membrane and enter the cell. But, alas, viruses and bacteria can evolve resistance to drugs. Bacteria can even exchange genetic information by lateral transfer of genes and so pass on resistance to antibacterial drugs to other bacteria. They do this by forming a bridge with another bacterium and passing on the DNA through this bridge – rather like very primitive sex. Improper use of antibiotics will help the infectious bacteria to evolve resistance, as if the dose is not maintained for long enough to kill them all, then those that have some mutations which make them a little resistant to the drug will be selected for survival, and further resistance can evolve. This is why it is important to continue to take the full course of antibiotics even if you begin to feel better: it is essential to kill all the invaders.

There are infectious organisms that are neither bacteria nor viruses. Malaria, an infectious disease widespread in tropical and subtropical regions, infects between 300 and 500 million people every year and causes between one and three million deaths annually, mostly among young children in sub-Saharan Africa. The disease is caused by a mosquito injecting a single-celled parasite, which lives in the mosquito gut, into the blood when it bites someone.

The parasites multiply within red blood cells, destroying them and causing symptoms that include anaemia, as well as other general symptoms such as fever, chills, nausea, flu-like illness and, in severe cases, coma and death.

A mutation in the gene for haemoglobin in our red cells can provide protection against malaria. This mutated gene has unusually high gene frequencies in populations where malaria is endemic; this is because individuals who have only one copy of the mutated gene, known as sickle-cell trait, have a low level of anaemia, but also have a greatly reduced chance of malaria infection. This mutation has emerged independently at least four times in malaria-endemic areas, demonstrating its evolutionary advantage in such affected regions. But individuals who have both maternal and paternal genes affected suffer from full sickle-cell disease due to the abnormal haemoglobin, and rarely live beyond adolescence. Another set of mutations found in the human genome associated with reduced malaria infection are those involved in causing blood disorders known as thalassaemias.

Another nasty human parasite lives in tsetse flies in Africa, and is transmitted to humans when the flies bite. This parasite, called a trypanosome, causes the disease known as trypanosomiasis, or sleeping sickness.

There was great concern in the United Kingdom in the 1980s when a number of people started to become seriously ill, and die, from a formerly very rare degenerative brain disease resembling the disease in cows called BSE or 'mad cow disease'. It was suspected that some of these cases at least were the result of eating BSE-infected meat. The agent that causes these conditions is quite remarkable – it is not a virus of the normal type, as it contains no nucleic acid, only protein. These types of agents are called prions, and not only cause terrible neurodegenerative illness but may also be involved in the long-term reproduction of blood stem cells. The prion protein is expressed as a normal harmless protein in many of our cells, and is believed to become infectious and propagate if it refolds abnormally into a structure able to convert normal molecules of the protein into the abnormally structured form that causes the damage.

Evolution is no fool, and our cells have acquired a brilliant defence system known as the adaptive immune response, which comes into action after the innate immune response. Its job is to destroy invading organisms like bacteria and viruses as well as any nasty toxins they produce. The system produces special proteins, antibodies, which can bind to foreign molecules like proteins and complex sugars, which are called antigens. The binding of antibodies to the antigens is an early step in their destruction and that of the cells to which the antigens are attached.

Our adaptive immune system which makes antibodies takes longer to act than the innate immune system but is the major defence against all the infections that surround us. It can save us from dying of an acute infection and, of great importance, often sets up immunity that protects us from any future infections with that agent. By itself, the innate immune system provides no specific immunity – but, like the police and army units that specialise in trying to keep out terrorists, evolution has given us a complex set of cooperating cells.

Because the cells of the adaptive immune system attack an enormous variety of foreign cells and molecules, they have to take special care not to attack normal ones in the body. They must always distinguish between foreign and self, but on occasion do not do so, and this causes autoimmune disease, as bodily organs are attacked by antibodies. There are illnesses that are due to the immune system attacking parts of the body, such as the joints in arthritis and the insulation of nerves in multiple sclerosis, as if they were foreign. But the ability of cells to make the distinction between self and non-self is mind-boggling, and it will show us cells at close to their cleverest. The system must also not attack helpful entrants like food, and useful bacteria like those in our gut.

One particular class of white blood cells, the lymphocytes, are responsible for adaptive immunity. There are as many lymphocytes in our body as there are nerve cells in our brain – billions. Their work begins only when the innate immune system is activated by invaders looking for a new comfortable and fertile home in our body. The lymphocytes of the adaptive immune system are able to recognize very specific features of particular infectious agents, and focus a concentrated attack upon the invader. There are two main types of lymphocyte. B lymphocytes, or B cells, produce antibodies, proteins that can recognise foreign antigens such as bacteria and viruses, bind to them, and tag them for destruction by other cells of the body. The T lymphocytes, or T cells, defend the body by killing infected cells or by provoking infected cells to rid themselves of their infection. They also 'help' B cells make antibodies.

The two classes of lymphocytes have their origin in

the blood-forming system, but they come from different organs. While immature B and T cells both start off in the bone marrow, T cells go to the thymus to continue their development and learn not to attack self, as those that do are eliminated. Just by looking at them it is not possible to tell which is a B or T cell even with an electron microscope. Both are activated only when they make contact with an antigen, but the B cells then secrete antibodies, while the T cells do not. T cells use a related but different type of protein to recognise antigens. Antibodies binding to a bacterial toxin inactivate it, and antibodies binding to a bacterium or a virus particle mark it for destruction by macrophages that will eat it up. T cells kill infected cells and can also activate B cells and macrophages. Foreign antigens also stimulate proliferation of the T cells to which they bind, which gives the body an immunological memory of invasion by particular bacteria or viruses – this is the basis of vaccination.

Each B cell produces just one type of antibody. The antibody molecule is inserted in the cell membrane, where it serves as the cell's receptor for recognising the corresponding antigen; then, when the B cell is activated by antigen binding to this receptor, large amounts of the antibody are manufactured and released from the cell. The simplest antibodies are Y-shaped molecules and the antigen can bind to the tips of the two arms of the Y. But antibodies are quite complex proteins with additional binding sites. The basic structure is made of four short protein-like chains, two heavy and two light, as they have fewer amino acids. The ends of these chains come together to form the antigen-binding site, and it is the variability of the amino acid sequence in these regions that give the binding site its specificity and variability.

It has been estimated that the immune system can give rise to a thousand million million different antibody molecules. Each type cannot be specified by a separate gene, as the number is way above the number of genes we have (just some 30,000). Instead, cells have invented a very clever mechanism that involves joining together a relatively small number of gene segments in different combinations to make different antibody proteins. For example, the so-called light chain of an antibody molecule is specified by a particular region on one of our chromosomes. Within this region there are sets of multiple copies of what are called V gene segments and J gene segments, which differ from each other. Before an antibody can be produced, one V segment, say V34, is combined with one J segment, say J21, to make a DNA sequence that codes for the variable part of the antibody light chain. A similar process produces a workable gene for the antibody heavy chain. This all occurs as a B lymphocyte is developing in the bone marrow, and once these arrangements have occurred they will not change – that lymphocyte will always produce antibodies with a V34 segment and a J21 segment in its antibody light chain. This process of rearrangement occurs in every developing lymphocyte, using different gene segments and producing a different antibody (or T-cell antigen receptor in the case of T lymphocytes) each time. Unlike normal cell differentiation, this process results in irreversible changes in the gene regions involved. The number of possible combinations of V and J segments can produce an enormous number – billions and billions – of different antibodies.

Combining different units to give different functions is, in general, a major and powerful technique used by cells.

As a result of this type of development, each newly made B cell will carry a different type of antibody on their surface. Lymphocytes are continually circulating in the blood and only a few will meet and recognise a particular antigen. When the B cell meets an antigen that binds to its antibody it proliferates to produce a large number of identical cells, which all differentiate into cellular factories dedicated to manufacturing and releasing that particular antibody into the blood. Thus one original B cell produces many cells to deal with the unwanted antigen-bearing invader. Some of these cells persist for long periods, so that if the antigen appears again at a later time there are lymphocytes ready to deal with it. This is the mechanism by which vaccination works: as the immune system 'remembers' meeting the antigen that characterises a particular infectious agent, such as a flu virus, and so has the right sort of lymphocytes already present when the real flu virus arrives.

T cells do not recognise bacteria and viruses directly. They recognise an antigen only when it has been processed by another body cell and is being displayed on that cell's surface, as is the case for those cells devoted to presenting foreign antigens on their surface. This happens, for example, in cells infected with viruses. T cells defend the body by either killing virus-infected cells or by activating other cells, such as macrophages and antibody-producing B cells, which have more direct effects on the pathogens. This requires the work of special antigen-presenting cells, which are key workers in this cooperative defence system and essential for activating killer T cells.

These antigen-presenting cells include macrophages, B cells and the most potent presenting cells, dendritic cells. Immature dendritic cells are present throughout the tissues of our body. When they meet an invader like a bacterium they engulf them or their products, and then go and find T lymphocytes to finish the job. Killer T cells are activated when the antigen displayed on the surface of an antigen-presenting cell binds to an antibody-like molecule on its membrane. But the antigen that is presented to the T cell must be partly degraded, and then carried on the presenting cell surface by special molecules called MHC. Once activated, killer T cells can kill infected cells displaying that antigen, or help other cells to do so. Note that the cells attacked will have the parasite, like a bacterium or virus, inside them, so that they are protected from antibodies such as those produced by B cells. The killing by the T cell is by inducing suicide - apoptosis - in the target cell.

T-helper cells are rather different from their killer co-workers: they do not kill, but when activated by antigen presenting cells, they can stimulate B cells to produce antibodies, can activate killer T cells, and can activate macrophages to destroy any evil fragments they have left behind. HIV is an RNA virus that causes acquired immunodeficiency syndrome (AIDS), a condition in humans in which the immune system begins to fail, leading to life-threatening opportunistic infections. This virus kills T-helper cells. There are only nine genes in the AIDS virus – a small number for all the problems it causes, one may think. When it enters a cell it directs the cell to make a DNA copy of its genes, which are then inserted among the genes of the host cell. There is a receptor for the virus on cells of the immune system, namely the T-helper cells, but also other cells like macrophages. The T-helper cells become depleted as the virus enters and destroys them, and when T-helper cell numbers decline below a critical level cell-mediated immunity is lost, and the body becomes progressively more susceptible to opportunistic infections and cancer. It has thus been very difficult to make a vaccine for this virus, which infects 1 per cent of the world's population. The problem has been to make an antibody which would block the virus from entering the T cells.

How does the adaptive immune system avoid attacking normal cells or proteins that are in our body? Fortunately there is an education programme in which the immune system learns how to distinguish foreign from self, and not to kill its own cells. This involves cell death when a molecule from the body acts like an antigen and binds to the lymphocyte. If cells are grafted between non-identical twins, the immune system will attack the cells, but will not do so in the case of identical twins. The learning process by which the immune system learns not to attack its own body occurs during development. Lymphocytes are attracted to the thymus, where those that recognise antigens in the body are eliminated, and this prevents autoimmunity – the production of antibodies which attack the body's own cells.

But as we have seen before, the best of evolutionary intentions can, alas, have very negative consequences. Autoimmune diseases are the result of the failure of the immune system to recognise some normal body cells. This can lead to the activation of T cells and antibodies to interact with normal cells and cause local inflammation and tissue damage. Nearly 10 per cent of the population is affected by autoimmune diseases, of which more than forty are known. These illnesses include rheumatic fever, rheumatoid arthritis, Type 1 diabetes, multiple sclerosis and the common skin disease psoriasis. Cells are aware of this autoimmune problem, and there is even one class of T cells that is devoted to repressing immune responses that could be harmful to the body, like attacks on healthy cells or our resident 'good' bacteria.

Rheumatoid arthritis is an autoimmune disease that mainly affects the joints in a very painful manner. It can lead to the destruction of the cartilage in our joints. The actual cause is not known, but there is a genetic susceptibility. Recent studies have pointed an accusing finger at the lining of the joint: in rheumatoid arthritis there is a dramatic increase in the number of cells in this lining layer and it becomes filled with cells that cause inflammation, such as macrophages, lymphocytes and mast cells. These cells secrete substances that, together with antibodies made against the joint cells, create the inflammation, and the actual cells of the lining themselves contribute to this damage. Type 1 diabetes results from the immune system destroying the cells in the pancreas that produce insulin, which enables glucose to enter our cells. Multiple sclerosis is the result of the immune system destroying the cells that provide the electrical insulating cover of many nerve cells, and so preventing the nerves from functioning properly.

We now turn to another failure of cells to behave properly: the terrorist activity of cancer cells.

12

How Cancer Strikes

how rogue cells form tumours

We are a society of cells that cooperate in wonderful ways to keep us alive and well – but cancer cells break all the rules of cooperation in this happy community. They try to take over the organism and have no concern for the neighbouring cells, which they even cause to die. Cancer is essentially the evolution of invading cells that compete with normal cells for space and resources. Just a single rogue cell from the billions of cells in our body can give rise to cancer.

Cancer has its origin when a single cell suffers an alteration in its genetic constitution that results in it continuing to divide when it should not, and so becoming a rogue cell. Misbehaving cancer cells grow and replicate with little restraint and give rise to a bunch of cancer cells, a tumour. A tumour may be benign, and have little ill effect on the local cells or our body. But even a small tumour in the brain can kill because of its local effects on the nervous tissue, which include compression and release of damaging molecules so that nerve cells no longer function properly. These effects can also occur in other tissues, and as the tumour enlarges it can block airways and blood vessels. In the gut, a tumour can lead to bleeding. A cancerous growth can have a variety of effects, and if it spreads the damage can be very severe. Further changes in the cancer cells can lead to an invasion of other regions of the body, which is almost always malignant and often fatal. Cancer is frightening and kills many. In the USA there is around a 40 per cent chance that an individual will have cancer at some stage in their life.

Cancer is not just one disease but many. There are more than a hundred different types of cancer, ranging from leukaemia to breast cancer. Cancer cells differ from normal cells in a number of ways. They have two defining characters: they continue to divide in defiance of constraints, and they invade foreign territories and destroy them. Cancer cells – unlike most normal cells, which can only divide a limited number of times – have limitless growth potential when cultured; they rely very little on external growth-promoting factors; they can evade suicide, which is induced in damaged cells; and they send out signals that direct blood vessels to grow towards the tumour and so provide it with oxygen and food. Cancer cells are not fast growing, and in general they divide more slowly than their normal counterparts, but they are more genetically unstable than normal cells and this instability can lead to greater malignancy.

With cancer, cell multiplication gets out of control due to an abnormality that arises in just a single cell. This abnormality is almost always due to an error in the cell's DNA. This is very clear in the case of leukaemia, a cancer of white blood cells where there is a crossover between chromosomes 9 and 22. All the cancer cells have this abnormality, although in different patients the crossover can vary. The initial cancer cell divides many times and gives rise to many more cells, which can become more malignant and invade other tissues. This is an example of cell evolution, because the most successful cancer cells are those that overcome the defences which our body has evolved to prevent their growth, and this allows them grow and invade. They behave like terrorists who try to take over the body. Their ability to do this needs to be understood.

Cancer may be the price we have to pay for having cells that continually repair and renew organs in our body, as most cancers arise from tissues that contain dividing cells, such as in the skin, lung and gut. Nerves do not divide, so brain tumours arise from the supporting cells in the brain and from nerve stem cells.

To form a tumour, the cancer must progress through many rounds of cell division and further mutations. The best way to think about the development of cancer is in terms of Darwinian evolution. As the descendants of that single rogue cell multiply, they are prone to have more mutations, and so there is positive selection for those cells that survive best, namely those that are most malignant. Many cancers, in addition to having mutations, become genetically unstable; they are unable to repair local DNA damage, and are even unable to keep a normal complement of chromosomes. These can increase in number, and also exchange regions, which can make them even more malignant. Individuals who lack DNA repair mechanisms will have a greater chance of developing cancer. One example are those people who lack the mechanism to repair damage caused by sunlight, and so get more skin cancers. In the end it is a case of mutated genes being too selfish, because when the cancer kills us the cancer cells die too.

Alterations in the DNA of the cell, and thus the absence or presence of key proteins as well as abnormal proteins, is the cause of nearly all cancers. It is now clear that most cancers arise from mutations in the cells' DNA or the rearrangement of its genes, but rarely from just a single mutation. This damage can be due to exposure to radiation or chemicals that damage DNA, or by faulty DNA replication. Lung cancer is largely produced by smoking, and there are common small changes in a gene that can increase the chance of cancer developing by 30 per cent. The time between exposure to a cancer-producing agent and the observable presence of a tumour can be as long as ten to twenty years, as is the case with smoking, and similarly with chemicals and radiation. However, not all cancers have their origin as a result of DNA damage. For example, some chemical agents can cause skin cancer by initiating cell multiplication in skin cells that were not dividing before.

Until some fifty years ago it used to be thought that viruses caused cancer by bringing in new genetic material which altered the cell's behaviour. A few cancers are caused by viruses, such as cervical cancer in women, and the AIDS virus can cause cancer by destroying the immune system to the extent that another virus, herpes, can infect the tissues and cause cancer. The viruses act by interfering with the host cell's genes and subverting the cell's control of cell division.

An interesting example is the Epstein-Barr virus, which causes Burkitt's lymphoma, the commonest childhood tumour in equatorial Africa. An important principle in science is that one should not give up one's beliefs at the first evidence that they are wrong. As Francis Crick pointed out, a theory that fits all the facts is bound to be wrong, as some of the facts will be wrong. A good example of this is related to the discovery of the cause of this cancer. In 1961 Tony Epstein, after whom the virus is named, attended a lecture by Denis Burkitt, a doctor working in Uganda, in which Burkitt described a peculiar cancer that he had observed. The tumour was thought to be caused by the bad environment in which the children lived. Epstein was, however, convinced it had a viral origin. He obtained samples of the tumour from Africa and his team hunted for the virus by trying first to grow the tumour lumps in culture, without success. But they went on and on, convinced that they were right. Then a sample arrived which looked a bit different. When Epstein examined it under the microscope he noted the single cells that had broken off, tumour cells, and which the researchers then cultured. When examined with the electron microscope a virus was identified which causes Burkitt's lymphoma.

Almost all cancers have their origin in a single abnormal cell, though by the time cancer is detected there will be billions of cancer cells derived from that single cell. And almost all cancers have a genetic basis and are caused by changes in our DNA. Yet there is a puzzle: it has been calculated that in our innumerable dividing cells, every gene will have undergone many mutations during our lifetime, so why is cancer so comparatively rare if it is due to mutations? The answer is that several specific genes, probably ten or more, need to be mutated to produce cancer, and the fact that this takes place over

181

years explains why cancer is a disease whose incidence increases with age.

There has been an extensive search for those genes which, when mutated and damaged, lead to cancerous growth, and some 350 cancer-related genes have been identified which are spread over all chromosomes other than the Y. These genes become responsible for human cancer if they are amplified, deleted or altered so that they make nonsense proteins. This research involves the sequencing of many genes, and the results can be complex and surprising. The sequencing of 13,000 genes from breast-cancer patients and from patients with gut cancer revealed mutations in nearly 200 different genes. There remains the problem of finding out which of these mutated genes are the root cause of the cancer. Worse still is the knowledge that mutations arise randomly in each case of each variety of cancer, and so each case of cancer will most likely have its own unique combination of mutated genes.

What is it that the cancer-causing genes actually do to the behaviour of the cell? Surprisingly, major clues come from the fact that most cancer-causing genes are already known to take part in the development of the embryo, and are thus involved in the social interactions between cells. Examples are those genes whose proteins are involved in signalling pathways that provide communication between cells. These genes are thus those controlling the social behaviour of cells. Sonic hedgehog is one such signalling protein involved in cancer. The main genes whose alteration causes cancer are those involved in the control of cell division in the cell cycle discussed earlier. For example, a gene that normally prevents entry into the start of DNA synthesis if the right conditions are not present, when mutated does not care about the right conditions and lets division begin.

Thus mutations can initially act by making the rate of cell division faster, and making the cell ignore normal constraints on cell division, such as preventing the cells differentiating into non-dividing cells. In the skin, stem cells divide to give rise to another stem cell and a cell that does not divide again; with cancer, the stem cell may continue to divide, or its sister cell continues to divide. It is a common feature of cancer cells that they fail to differentiate properly. This is particularly true of cancer cells in sheets like the skin, where there is a continuous replacement of cells. This failure to differentiate is also seen in leukaemias, or cancers of white blood cells. The cells have become stuck at an early, immature stage in their normal development.

There is evidence that in a tumour, stem cells are the driving force, and also that stem cells are often the original source of the tumour. Stem cells are present in leukaemia, and this fits in with all blood cells coming from stem cells. Most of the cells in a population of mouse leukaemia cells cannot give rise to the tumour in another mouse when they are injected; only stem cells in the tumour can do this, and they are just a small proportion of the tumour. With leukaemia there may be a special niche in the bone marrow where the malignant cells are provided with the molecules that support the tumour. There is also evidence for stem cells in breast tumours. It is not hard to see how stem cells could give rise to tumours as they are the only cells in our body with an apparently 'infinite' capacity for division that could escape from the local controls keeping them in check. The stem cells in a tumour present a problem for treatment, as they may not be dividing when the tumour is treated with radiation or drugs, and will thus survive. They also seem better able to repair damage to their DNA than the cells to which they give rise.

There are, surprisingly, some genes that can cause cancer when they become overactive. Just one copy of such a gene, when mutated to be active, can cause cancer, and such a gene is known as an oncogene. More than a hundred oncogenes have been identified. Some cancers are caused by the cells producing growth factors and responding to them. A new way of treating such cancers could be based on using an antibody against the receptor for the growth factor. There are also what are known as tumour suppressor genes, which can cause cancer when they are inactivated. These genes can cause cancer if they are damaged and no longer function, and so are effectively lost; an example is the gene which causes cancer of the gut if both copies of the gene, one from the mother and the other from the father, are made inactive.

The gene BRCA-1 is associated with breast cancer – about 5 per cent of the 44,000 cases each year in the UK are linked to this gene. Prenatal diagnosis that selects embryos that do not have this gene is currently being used, and elimination of such embryos means that the mother will not bear a child that will have a gene linked to breast cancer.

What has evolution given cells to protect us from cancer? The p53 gene has been called the guardian of the genome. Activation of p53 when things are going wrong, and the cell may begin misbehaving in a way that could lead to cancer, causes cell-cycle arrest, as well as cell death by suicide. It is thus essential to prevent p53 becoming activated when things are going well. There are fortunately many regulatory proteins that control activation of the gene and provide the necessary safety measures. Activation of p53 is due to stress on the cell, the major ones being damage to DNA and oncogene activation. If the p53 gene is mutated, the chances of cancer developing are greatly increased. Lifesaving as p53 can be in preventing tumour formation, it also has, unfortunately, less beneficial effects. It will not be able to block the formation of all tumours, and when a tumour is treated by radiation or anticancer drugs, the damage this causes to normal cells may activate p53 and cause the death of the normal cells. It could thus be helpful to turn off p53 when giving chemotherapy for tumours. On the other hand, it would be very good if p53 could be specifically activated in the cancer cells. Very recent work has shown that in mice restoring the activity of p53 in a tumour can halt the growth of a tumour, and even cause it to disappear. Thus one Holy Grail is to find a small molecule that will reactivate p53 in a tumour. What complex lives cells live!

The continuing division of cancer cells may be related to the presence of the repetitive DNA sequences at the end of their chromosomes known as telomeres. In most cells, these telomeres shorten at every cell division as the enzyme telomerase is not there to lengthen them, and when they are virtually absent the cell is no longer able to divide. This may explain why cells in culture have a limited ability to multiply if they cannot restore telomere length after each division. By contrast, cancer cells do make telomerase, which restores the length of the telomere after every cell division. It may be that some cancers come from cells that have lost their telomeres but have then acquired the capacity to make telomerase again, which lets the telomeres lengthen again and the cells divide.

More complexity comes from the involvement of micro-RNA genes in cancer. These, as explained in Chapter 5, do not code for proteins but produce a strand of RNA that can destroy a messenger RNA so that the protein it codes for cannot be made. The micro-RNAs are largely absent in some forms of leukaemia. Micro-RNAs repeatedly pop up in the analysis of the chromosomal regions involved in cancer, but their role in cancer remains unclear.

The invasion by cancer cells from the original tumour to other sites is known as metastasis and is one of the most lethal activities of cancer cells. When they arrive at foreign sites some distance from their origin, they compete with the local tissues for food and oxygen and damage the local cells. Even small tumours can shed a million cells a day, and some of these seem determined to invade. They secrete enzymes that break down the tissues surrounding them and so can reach blood vessels and even attract blood vessels to them. Tumours can spread to many different parts of the body by such means.

In order to move away from the original tumour, the cells must lose their tight contact with their neighbours. This probably requires destruction of the adhesion molecules that hold them together. In addition, if they have their origin in a sheet of cells, they are not normally cells that move or migrate, and so to be invasive they have to change their character and become cells that migrate and change their shape. Cancer cells can be invasive because their movement is not inhibited by neighbouring cells. In culture, normal connective-tissue cells have their movement inhibited when they contact another cell – this is called contact inhibition. When a normal cell is crawling over the dish and it meets another cell it stops sending out extensions at the front end, and so cannot cross over the other cell. Cancer cells are not inhibited in this way, and this helps them to invade and spread to distant sites. Once they have moved into a blood vessel, they can spread to other sites far from their origin. But when they reach a blood capillary, they must again cross the vessel wall to infect the local tissue. There are barriers to this long-distance voyage, but there are, alas, many cells trying to make the voyage, and many are successful.

A special type of cancer arises without any mutations. Teratocarcinomas can arise spontaneously from germ cells. The tumours are quite bizarre, as they contain a mixture of differentiated cell types. They are very relevant in relation to embryonic stem cells, as if these stem cells are placed under the skin of a mouse they develop as teratocarcinomas. Thus embryonic stem cells have a quite strong ability to form this type of cancer.

Cancers of the breast, lungs and intestine are among the most common. Skin is also a common site for the type of cancer that humans suffer from, and there are several million cases worldwide each year. A particular form is melanoma, a cancer that comes from the pigment cells in our skin and that accounts for most skin-cancer deaths. These pigment cells have their developmental origin in the neural crest quite early in development (see Chapter 7) and migrate into the skin, where they have a key role in protecting our skin from the sun's radiation and help to prevent skin cancer. But – another cell irony – cancer can develop from these pigment cells. Mutations in pigment cells can lead to the breakdown of their junctions with neighbouring cells; escaping their tight containment by other skin cells, the pigment cells multiply and spread, and can give rise initially to moles and then to cancer cells that can spread through the skin and into the body.

When tissue from an unusual lump in our body is isolated and examined under the microscope, in cases of malignancy the individual cells have an abnormal appearance and their arrangement is disturbed. In addition the tongues of invasive cancer cells infiltrate surrounding tissues, while benign growths have a smooth border. Pathologists can judge the seriousness of the growth based on such characteristics, particularly whether it is invasive, which occurs in only a few tumours. The tumour, however, is not easily diagnosed until it contains hundreds of millions of cells.

Cancer, like all cell communities, is full of surprises. While normal cells rely on oxygen to make the ATP for energy, cancer cells also break down glucose without oxygen, and this gives them the advantage of not being dependent on a supply of oxygen. There is some evidence that the inflammatory response, which was considered previously, can support cancer and even assist the cancer developing. It is the innate immune system that is responsible, not the more advanced adaptive immune system whose antibodies cancer workers have tried so hard to harness in the fight against the condition. The inflammatory process is involved in the middle stages of cancer development and the cells treat the cancer as an organ to be supported; cancer is even viewed by some as a wound that does not heal. One of the cell types involved is macrophages, which scour the body to eat up unwelcome visitors and can kill cancer cells. More than half of a cancer mass can be made up of non-cancer supporting cells such as fibroblasts and macrophages, which are part of an inflammatory response and needed by the cancer cells. Cancer cells can even re-educate macrophages to produce factors that promote the growth of the tumour. Even worse, the substances produced by the inflammatory cells can encourage the cancer cell to migrate to other sites. But all this could be turned around by engineering the macrophages to deliver a virus to the tumour; when released, the viruses would kill the cancer cells.

There are many other errors that cells make, and we now look at some of these.

13

How Diseases Are Caused

when cells behave abnormally

Clever and reliable as our cells are, things can sometimes go wrong. Cells lead moderately stable lives and adapt to small natural variations in their environment, but they can also be damaged, and many human illnesses have their basis in cells being damaged or misbehaving. We have already looked at the effects of infectious agents like bacteria and viruses, and at cancer, and now we turn to many of our other illnesses that are due to a variety of mistakes in the society of cells. Many errors are due to mutations in genes affecting cell functions in different parts of the body; they can result in illnesses ranging from cystic fibrosis to muscular dystrophy. But the causal chain from mutation to illness is only fully understood for a few illnesses.

In any organisation the errors made by just a small proportion of the members can have effects on the whole community. Errors in the supply of blood from the factory in the bone marrow and its distribution via the blood vessels provide many examples. Damage in one region of the circulatory system can cause damage in others. Cells need the food and oxygen that our circulation provides, but there can be a lack of red cells and leaks and blockages in the system, with severe effects. Too many fat cells lead to obesity, and a lack of insulin can lead to diabetes. Even damage to the nuclear membrane of a cell can have bad effects.

Cells can be damaged by a loss of blood supply. Red cells carry oxygen on their haemoglobin to all our cells; reduction in this essential transport can result in anaemia, causing a feeling of fatigue in general or during exercise. People with more severe anaemia can have shortness of breath; severe anaemia increases cardiac output, leading to palpitations and sweatiness, and even to heart failure. Anaemia can be the result of blood stem-cell failure and can also result from iron deficiency, as this causes the red cells that are produced to be smaller and to contain little haemoglobin. Many fewer functional red cells in the blood can lead to widespread organ dysfunction, and one must remember that normal blood cells only remain in circulation for about a hundred days before they are destroyed. Loss of blood supply reduces the production of ATP by the mitochondria because of the lack of oxygen. If this occurs in the heart muscle the cells cannot contract properly, which affects heart pumping. The lack of ATP also results in the sodium pump in the cell membrane functioning at a reduced level and sodium ions and water entering the cell, causing it to swell; this swelling disorganises other cell functions such as protein synthesis. These changes are reversible if oxygen is soon supplied again, provided that the mitochondria and surface membrane have not been too badly damaged.

Errors in the red cells themselves, particularly in their haemoglobin, can also lead effectively to loss of oxygen. Sickle-cell disease results in red cells taking on a sickle-like shape because of their abnormal haemoglobin; they become less flexible than normal red cells and have difficulty being squeezed through small capillaries. The cause of the change in shape is a mutation in the haemoglobin gene that results in a single amino acid in the haemoglobin being changed; the haemoglobins then fold in a quite different way and form a lumpy structure that deforms the red cell. This is one of the clearest examples of how a mutation can exert a recognisable effect on an organism, as there is not a complex long sequence of interactions between the mutation and the observed effect. As described earlier, sickle-cell disease is most common in regions of malarial infection, where a mutation in just one copy of a haemoglobin gene helps protect against malaria.

As in the network of pumps and pipework in the plumbing system of a house, things can go wrong with the circulation. When the circulatory system in our body, which includes the pumping heart and the numerous vessels, arteries, capillaries and veins, goes wrong it can result in damage that ranges from heart attacks to strokes, but we have cellular plumbers who try to prevent such failures.

Clotting of the blood is one such plumbing activity. While it is essential for stopping bleeding from a damaged vessel, it is also essential that clotting does not occur in normal circumstances, as this can block a blood vessel and have, as we shall see, disastrous consequences. Normal clotting in a damaged vessel results from platelets adhering to the damaged area; they then release factors that together with the platelets form a fibrous plug to stop the bleeding. Platelets are small disc-shaped fragments of cells produced in the bone marrow. Their sole function is to prevent blood loss. At the site of damage they change from a disc into a sphere and put out long, fine extensions over the damaged surface. These small plumbers only live for about a week in our circulatory system.

Strokes, which are a major cause of death and disability, arise precisely because our body cells do all they can to avoid blood loss by forming a clot at the site of injury, ensuring that the blood retains its fluidity and can flow through the arteries, capillaries and veins. But if a blood vessel is damaged, then at the site of damage the blood can be converted from a fluid to a solid state made up mainly of platelets and fibres as in a clot, and this can block a key blood vessel to the brain and cause a stroke.

This clot is known as a thrombus. It can cause a stroke if it blocks blood flow to the brain and so causes damage to the brain's nerve cells. The damage that gives rise to a thrombus may have its origin in physical causes, or through infection or surgery. A thrombus can develop in the arteries if the lining of the vessel is damaged so that platelets come in contact with the underlying cells, or if the blood flow becomes turbulent. Formation of a thrombus is more common in veins, particularly in the legs of someone who has been immobile for some time – as on a long plane trip. But if the thrombus blocks blood flow there is fortunately, in some cases, a way the blood can avoid this traffic jam by taking other minor routes through other vessels nearby.

The danger of thrombi increases if they can become emboli, clots which can move and can be carried in the blood from one area in the circulation to another, so that they block the circulation at this new site. Emboli originating in veins can be moved by the blood to the right side of the heart and end up in the lungs. If the embolus is large and blocks a large lung artery, then the blood pressure will drop dramatically and can lead to instant death. Smaller emboli may just damage the lung. Emboli originating in arteries can end up at the brain, and the loss of blood circulation in this region can lead to the death of many nerve cells, and so to the death of the individual.

Our blood vessels can become furred up with fats, and as they become narrower the formation of a thrombus becomes more likely. Atherosclerosis is the condition that results from the lesions in blood vessels blocking the normal flow of blood because of fibrous fatty plaques. The plaques develop mainly in arteries and can be considered to be a chronic inflammatory response to injury of the vessel lining, but the cause of the injury is not known. The plaques protrude into the vessel and may also have their origin in local injury combined with a high concentration of fatty molecules. Our hard-working heart muscle is very dependent on a good blood supply to give it oxygen and nutrients. If the vessels to the heart are too narrow, poor blood flow results in heart muscle dying and so to heart attacks, which can be fatal.

Osteoporosis, a disease of bone, can be caused by a reduced blood supply activating cells called osteoclasts that remove bone. Their normal function is connected to bone remodelling as we grow, but if there is a reduction in oxygen in the bone because of poor circulation, osteoclasts remove bone instead, which makes the bone more likely to break.

The sugar glucose is the main source of energy for cells, particularly for muscle. But without insulin, as pointed out earlier, glucose cannot cross the cell membrane and enter the cell. Insulin is produced by beta cells in the pancreas, and Type 1 diabetes results if these cells stop producing insulin. Type 1 diabetes is an autoimmune disease in which the beta cells are mistakenly attacked and destroyed by the immune system, and this condition usually becomes apparent at quite a young age. Type 2 diabetes, on the other hand, which is more typically a disease of later life, is due to a failure of receptors in the cell membrane to respond to insulin – in other words, cells have become insulin-resistant. Obesity is a major risk factor for diabetes Type 2 because the enlarged fat cells release increased amounts of different substances including particular fatty molecules that cause insulin resistance. By contrast, lean fat cells promote glucose uptake. In both types of diabetes, glucose uptake by cells is greatly reduced, which lowers their ability to produce energy and leads to serious long-term complications, including cardiovascular disease, kidney failure, blindness, nerve damage, and blood-vessel damage which may cause impotence and poor healing. Poor healing of wounds, particularly of the feet, can lead to gangrene, which may necessitate amputation.

What about errors in genes that cause illness? Mutations in genes are a major cause of many illnesses from muscular dystrophy to depression because proteins are altered, absent, or present in the wrong cells. My reason for giving them rather little attention so far is that in most cases where a genetic basis has been identified, it is very difficult to understand how the genes affected have caused the alterations in cells that lead to the illness. There is a nearly 70 per cent chance of a disorder due to faulty gene action affecting an individual's health sometime in their lifetime – heart disease and cancer are the most common. Some 50 per cent of spontaneous abortions are due to genetic abnormalities, and 1 per cent of all newborn babies have a genetic abnormality. Substitution of just one nucleotide base by another in the millions we have in our DNA is a point mutation, and can result in an altered protein and so possibly some abnormality. Having your genes sequenced at this stage in our knowledge could give you only a small indication of what genetic illnesses you may acquire.

While we have only some 30,000 genes, alternative splicing of the messenger RNA results in our having some 100,000 different proteins, in which there may be mistakes that alter their function. The most common form of variation in our DNA is the alteration in just one nucleotide out of the many millions. There can be very many of these in different regions, and a few will be in the coding regions of a gene and so can give rise to an illness such as sickle-cell anaemia. Alterations in non-coding regions which may be involved in the control of gene activity can also have bad effects. For example, a mutation in the control region of a gene involved in cholesterol synthesis in the liver results in high levels of cholesterol in the blood, with increased risk of circulatory and heart problems. Each of us has mistakes in our DNA which, hopefully, do not have any obvious negative effects by causing the production of bad proteins or absence of good ones.

The causal sequence from an altered or absent protein to an illness can be long and complicated, and is usually not well understood. An even bigger problem is that for most illnesses there are a number of genes, sometimes many, involved, and so how the combination of mutated genes gives rise to the cellular basis of the illness is hard to trace. Recent studies have identified 24 genetic risk factors for common diseases such as diabetes, bipolar disorder, inflammatory bowel disease and arthritis. This work is based on the examination of 17,000 individuals. That is why sickle-cell anaemia is such a good and clear example, as we can understand the causal sequence.

Nevertheless, there are other examples of genetic diseases that at least re-emphasise how abnormal genes can affect our lives. Muscular dystrophy is an inherited disorder that results in muscle weakness when the muscle fibres are replaced by fat. The gene involved codes for a protein that is found close to the membrane of the muscle, but it is not clear why its absence has such bad effects. Cystic fibrosis is a disease associated with membrane dysfunction. It is due to a mutation in a gene that codes for a protein channel that lets chloride in and out of cells, and its harmful effects are mostly concentrated in the pancreas and the lungs. The protein is synthesised, but the mutation causes it to fail to move to the Golgi apparatus of cells, and so it fails to get to the membrane. A dense mucus forms on the cell surface, which cannot be cleared, leading to lung and pancreas malfunction. Deletion of a particular region of DNA can lead to Prader-Willi syndrome, which results in mental retardation and is due to faulty imprinting, as explained earlier, on chromosome 15. Children who have a mutation in both of the genes that code for the enzyme that acts on an amino acid can have mental retardation if they eat much protein, an illness known as phenylketonuria. Infants are routinely screened for this at birth so that any affected can be put on an appropriate diet.

A mutation in the receptor for the male hormone testosterone makes the cells of a male insensitive to testosterone and will result in that XY male developing all the external features of a female. Internally, the testis will begin to develop. There are also abnormalities with the number of sex chromosomes. Klinefelter syndrome is a cause of male infertility where the male has an extra X chromosome. Turner syndrome in females results in a failure to develop normal female sexual characteristics and is the result of having just one X chromosome instead of the normal two. Mutations that occur in the X chromosome mostly affect males, as they only have one X compared to females who have two. Then there are inherited disorders due to alteration in the numbers of other chromosomes. Down's syndrome, a major cause of learning disabilities, is due to the presence of an extra chromosome 21 in the cells – something went wrong at meiosis. There is still little understanding as to why that extra chromosome has the effects it has.

Mental illness can be partly caused by genes. Many of the genetic effects are due to multiple gene mutations affecting the brain, as well as due to life experiences. These can be complex, and identifying the genes involved is difficult. Identical twins have the same genes, and if one has a severe depression the other one will have a 50 per cent chance of having a similar illness. For bipolar depression the chance is 75 per cent. Schizophrenia also has a strong genetic basis and there are claims that worldwide there is a 1 per cent incidence of the condition.

It has been possible to identify one of the many genes linked to depression, one that is involved in blocking serotonin uptake, but we are still a long way from understanding why this leads to depression even if there are plausible theories. It is thought that low serotonin is one cause of depression. Serotonin is widely distributed in the body and is the neurotransmitter at many synapses in the brain. There are special mechanisms in nerves for the uptake of the serotonin that is present around the cells, and antidepressants, like Prozac, are thought to act by blocking this uptake and increasing the local concentration of serotonin. Mutations in the gene coding for the receptor of the neurotransmitter dopamine can apparently lead to novelty-seeking behaviour. It is on such subtle processes going on in our brain cells that our mental health may depend.

Most mutations that have an effect are due to loss of normal function by the protein affected, but some mutations can result in the protein acquiring a new but unwanted function. An example of complex human behaviour due to a gene defect leading to a new function for the protein is Huntington's disease, a degenerative disorder of the nervous system which causes uncontrolled movements, loss of intellectual faculties and emotional disturbance. In this disease the mutated protein is toxic to nerve cells, which die in certain regions of the brain and result in the death of nerves that regulate voluntary movement. Huntington's is a late-onset disease – most of the other severe single-gene diseases show up in children

– and so those with the condition do not know they have it at the time they may have children. The mutation in the gene on either the male or female chromosome has the effect, so there is thus a one in two chance of a child inheriting the disease from a diseased parent.

Another brain error that affects the nervous system of many people is Parkinson's disease. This is a neurodegenerative disorder that results in muscle rigidity, tremor and a slowing of physical movement. The cause is the loss of nerve cells in the brain that use dopamine as their neurotransmitter. The loss of these nerve cells is thought to have a genetic basis, but to be due not to a classical gene defect but to the absence of particular micro-RNAs, and it is not an inherited illness. Micro-RNAs are being more and more implicated in both developmental and disease processes, and may also be involved in Alzheimer's disease. They function by causing the destruction of the messenger RNA of specific genes so that the protein they code for cannot be synthesised.

Gene therapy is the insertion of genes into an individual's cells to treat a disease, particularly those caused by abnormalities in the genes. There was great optimism some thirty years go that it would be possible to use this technique to treat, for example, cystic fibrosis, but it has turned out to be very difficult to properly insert a normal gene into the cells which lack it, and to ensure that the inserted gene acts normally. However, there has been a recent encouraging advance in gene therapy: inserting a gene into the retinal cells of individuals suffering from a genetic abnormality that leads to blindness has led to improvement in their sight.

*

Having looked at many illnesses from the viewpoint of the cells behaving abnormally, it is necessary to consider the views of those who invoke a quite different approach to illness. Practitioners of alternative medicine propose all sorts of ideas about illness which make no sense in relation to what we know about cells. I would certainly join a Society for the Protection of Scientific Terms. The urge was particularly strong after I had experienced Shiatsu, which is like acupuncture without needles, not unpleasant. But I was then told that I had very low kidney energy. The idea that pounding and pressing bits of my body could detect such a localised quantity of energy was irrational, absurd, and finally infuriating.

Science is supposed to use words with care and preferably with strict definitions. Non-scientists have taken a scientific term and used it in a way that seems to be totally inappropriate; but because the word is from science it gives it a spurious validity. Nowhere is this more evident than with the widespread use of the term 'energy' in what is politely called alternative or complementary medicine, but which bears little or no relation to science-based medicine. Thus Ayurvedic medicine claims that there are canals in the body carrying energy, and *qi* energy channels are central to acupuncture; crystal healing is based on transmission of energy, and faith healing also works, it is claimed, by channelling energy. There is no indication of how this energy is generated or what its nature is. Positive results are most likely due to the placebo effect. All this bears no believable relation to the fundamental concept of energy used in science. Energy is the capacity for doing work, as recognised by Galileo. Lifting a weight by a pulley requires a force moving through a distance;

the product of the force and the distance is work and is equivalent to energy. The unit for energy is the Joule, which is defined as the force of one Newton acting over one meter. All energy is associated with motion – the lifted weight, for example, has potential energy for moving down, when it is converted into kinetic energy. There is also chemical energy in the movement of molecules and thermal energy from heat. Energy is never lost, merely converted into another form. The main source of energy in our world is radiation from the sun produced by thermonuclear reactions.

Cells use energy to maintain their complex organisation and to grow, multiply and move, and we have seen how this energy comes from ATP. ATP provides energy when one of the three phosphates is removed. It is ATP that provides the energy for the function of our liver, kidneys and brain – in fact, for all our organs. There is no evidence whatsoever for a chemical or biological entity such as *qi* or kidney energy.

Such misuse of a scientific term does matter, for it gives a false credibility to the explanations offered for the possible success of how treatments like acupuncture might work. I am not suggesting here that the techniques of alternative medicine do not help patients or that their effect is entirely due to the placebo effect, but it is totally misleading to claim that energy underlies any possible effect. The continued misuse of the term should make one suspicious of those who use it in this way.

14

The Origins of Life

the mystery of the first cell

Darwinian theory makes clear that we have evolved from the first cell over billions of years. How did that first cell evolve? What is the origin of life? How single cells evolved into multicellular organisms is a much simpler problem than the origin of the first cells – once cells evolved it was basically downhill all the way to we humans. We start first with the evolution of nucleated cells from bacteria which evolved into mitochondria, and then move to the origin of multicellularity, and finally return to the great question, life's origin.

The framework for understanding evolution is Darwin's theory of natural selection. Evolution selects those cells that reproduce best, and thus selects those whose genes code for proteins that enable the cells to survive best. This is essentially Richard Dawkins's concept of the selfish gene.

Studies on rocks show evidence of life that dates back some three to four billion years. Cells have evolved by Darwinian evolution, and this undoubtedly required much time. There are now more than 10 million different types of living organisms on our planet, all composed of cells. They have all evolved by natural selection: that is, they reproduce and there are occasional variants that turn out to be more successful and so survive better, while others die out. This process can be very clearly seen in groups of cells. A good example, though bad for us, is cancer. The cells of all animals and plants have evolved from much smaller single cell organisms such as bacteria and other single-cell organisms. One of the cleverest of cells' abilities was to form multicellular societies. This was, as will be suggested, probably an initial advantage because the cells could eat each other in hard times. The great unsolved problem is the origin of the first cell, life itself. Is it more than just the lucky outcome of complicated chemistry?

Evolution of cells depends on changes in the genes, for they are the carriers of inheritance from one generation to the next. Changes in genes lead to changes in proteins and thus how cells behave; if the resulting behaviour is more successful then evolution will select it and the others will die out. It is changes in the coding regions as well as the control regions that can bring about such changes. Genes can have extra copies made. This provides new material for evolution, as there is now the original gene or genes and also new ones with which evolution can tinker. Tinkering, or fiddling around with what is there, is the main mechanism of evolution. A nice example is exon shuffling, in which two different genes can be broken up and then the segments joined up to make a new gene. Evolution proceeds by small steps; there are rarely any big jumps.

Evolution by selection is an astonishingly powerful process for generating complex functioning systems. The process can be illustrated by genetic algorithms that are used to solve technical problems on a computer and are derived from Darwin's ideas about natural selection and evolution. The basic idea is to use the computer to generate, randomly, a number of solutions to a particular problem that you want to solve. You then look and see which have provided a hint of a solution to your problem, and you select from these the fittest, eliminating the others. You then let these fitter solutions mate, as it were, by exchanging information, and so breed, and also introduce some mutations. The procedure is repeated, and again those with the highest fitness are repeatedly selected. Repeating the procedure again and again can give novel solutions to the initial problem.

Adrian Thompson from the University of Sussex has used this approach to design an electronic microprocessor that can perform a simple task like responding appropriately to the spoken commands 'Go' and 'Stop'. Instead of designing the circuit in the standard way, using the on/off switching of computer technology, he wanted to exploit all the properties of the on/off processes used by computers in the same way that biological systems exploit, for example, the chemistry of proteins. The approach worked remarkably well and the processor that evolved could detect subtle differences between Go and Stop. But how did the processor actually work? Thompson did not initially understand, and he had to work hard to find out. The system had, for example, evolved a clock, and it was complicated because it used all the resources of his machine. Thompson had not designed the circuit, but had helped it evolve by selecting circuits that looked as if they had some of the features he required.

Clearly this is a very powerful approach, and cells have used a similar one to arrive at organisms like us. Evolution of a process results in mechanisms that would not have been designed by an engineer. Evolution cares only about success, and this explains many of the complicated processes in cells, like those weird transmissions of signals from the membrane to the nucleus, illustrated by the Goldberg cartoon in which there are strange sequences of protein interactions from the membrane to the nucleus when a signal arrives at the membrane.

Bacteria, cells without a nucleus or mitochondria, are the most diverse organisms in the world. They had their origin some three billion years ago. Then some of them, at some stage, evolved into more complex and much larger cells with a nucleus containing the genes, and mitochondria for making energy. These cells no longer had a tough cell wall and could rapidly change their shape in a variety of ways. Just how such cells evolved from bacteria is not known, but it is clear that mitochondria evolved from a cell engulfing a bacterium, which then retained some of its functions and became a mitochondrion with its own DNA and the ability to replicate in the cell and provide it with energy. Perhaps the engulfing cell did not use oxygen to get its energy, but the mitochondrion that evolved from the engulfed bacterium could do that very well. Later, some of the genes for the bacterium were transferred to the cell's DNA in the nucleus.

The single-celled organisms that populated the earth some billion years ago had nuclei and mitochondria and were doing well, and evolved into many different single-celled organisms such as amoebae. They were both numerous and varied, and reproduced happily by cell division. So why did multicellular organisms evolve? What was the advantage of being a society of cells? The standard answer is that over time it provided the possibility for division of labour amongst the cells so that some could become specialised for particular tasks such as digesting food. But this does not explain how or why they became multicellular in the first place. Imagine a scenario in which one of those cells undergoes a mutation so that when it divides into two, the daughter cells remain together, and with further divisions continue to stay together and form a multicellular colony. When the colony gets too big it just fragments, and the cells continue to grow again into a new colony. The selective advantage promoting the survival of this organism was that when food was in short supply, and single cells were dying all around, some of the cells in the colony could eat their neighbours and thus survive. This might be the origin of multicellularity. Later on, one of the cells could have been selected to be fed by the other cells and then give rise to the colony, and so it became the egg.

There are, in fact, current examples of this process in animals like hydra and flat worms, which when starved maintain their smaller form by some cells eating others. It is not too difficult to imagine that certain cells near the centre of the primitive colony became specialised for eating their neighbours and began to do so even when times were not bad. These cells are the origin of the egg, for they could eat and grow large and then divide to give a number of cells that could form a colony. It is striking that the sponge egg cannibalises its neighbours to get bigger, and that all eggs are fed by other cells in the organism. In all animals, one can think of the egg as the cell for which all the other cells sacrifice themselves – it is the only cell that survives, because it gives rise to another animal. One of the major functions of all the cells of the female body is to ensure that the egg is fed. So our origins may lie in cannibalistic altruism or altruistic suicide a long time ago. Sex and the origin of sperm came later. But we are left with the question of the origin of the first cell, the origin of life. It was once thought that there was a special force in the cosmos that led to spontaneous generation of life. This belief goes back to Aristotle, who was a wonderful thinker but wrong about almost all the science he wrote about. This was a view still held a few centuries ago, but the invention of the microscope made a real change.

Following several experiments which were initially claimed to favour spontaneous generation of life, the theory was disproved in 1859 when Louis Pasteur boiled meat broth in a flask and bent the neck into an S-shape so that air could get in, but not bacteria. No bacteria developed in the broth; but when he tilted the flask so that bacteria could enter from the air, the broth became filled with growing bugs. It was realised that the bacteria entered from the surrounding air, and spontaneous generation was rejected. The presence of bacteria in the air is also of great clinical importance. So then there had to be a search for the origin of cells way back in time.

The origin of the first cell, life itself, was probably due to an emergent process or event that involved some sort of self-organisation, rather like the formation of little bubbles from a soap solution. It is not that easy to see why such bubbles form spontaneously. To create life there also must have been many interactions between molecules that formed such bubble-like vesicles, but also crucially involved the use of energy and the evolution of self-replication.

The oldest rocks to have an indication of life, as shown by the presence of a form of carbon found only in cells, are about four billion years old. There have been claims for life on Mars, but these have been rejected. Researchers have struggled with this problem: how cells could have emerged so long ago from our then barren planet, which had been formed some billion years earlier. By life, we mean a system that multiplies, has heredity, and that has offspring related to the parents but which are also quite varied. All these are essential for Darwinian evolution, and cells have all such properties.

The original building blocks might have had their origin in electrical activity in the atmosphere. For example, some of the contents of rocks like pyrite found near volcanic vents can convert carbon dioxide into compounds such as amino acids, which are essential components of the proteins on which much of life is based. There is also some evidence that the action of sunlight on certain minerals could give rise to the precursors of RNA and DNA. The earliest cells – bacteria – did not use oxygen, as there was hardly any free oxygen around in the atmosphere. The cells used photosynthesis and the breakdown of other compounds until some two billion years ago, when oxygen came into the atmosphere.

In that ancient and primitive landscape, the raw materials from which cells evolved were water, carbon dioxide, nitrogen and perhaps a little ammonia – not a promising collection or a pleasant environment. The key molecules in the origin of life, and in life now, are remarkably few. There are nucleic acids, DNA and RNA with their long strings of nucleotides, proteins with their long strings of amino acids, and two key types of sugar for energy. There are also some fatty molecules.

Several scientists in the 1950s thought about how life could have come from our earth's rather boring sea with the help of the sun. The sun provided intense radiation and temperatures varied greatly from the surface of the sea to miles deep down. Of particular importance was the suggestion by the chemist and Nobel Laureate Harold Urey that life might have started by making organic molecules in that primitive atmosphere. One of his students, Stanley Miller, was having a difficult time with his research project and he asked Urey if he could try an experiment to create amino acids from the components of that atmosphere. With reluctance, Urey agreed.

Miller then designed a simple apparatus that linked a vessel containing heated water to a supply of a gas containing methane, ammonia and hydrogen, and the gas was exposed to two metal electrodes which sparked to mimic lightning. Within a few days, the water had turned yellow and there was black material at the bottom. Examination revealed that he had created some half dozen amino acids, the key components of proteins, as well as some other molecules. The publication of the results in 1953 was a bombshell, and it is a curious coincidence that this is the same year that the structure of DNA was discovered. Since then, other key life molecules, like energy-giving sugars, have been formed by using similar methods. One important change was to do the experiment at low temperatures, and after twenty years this gave rise to molecules related to nucleic acids.

Exciting and encouraging as these results were, since they showed that at least some key molecules could be made in those early conditions, many problems still remained. Larger molecules, like proteins and nucleic acids like DNA, still needed to be made. There was also the nasty fact that lightning and the sun would tend to fragment molecules rather than make them larger. And while the sun is essential for the energy of life, the deep sunless oceans may also have played a puzzling but key role.

A new view of the origin of life opened up in 1977 when geologists were investigating the underwater terrain of the sea near the Galapagos Islands. They were one and a half miles beneath the surface in a region of volcanic activity, where hot water jets into the cool surrounding water. There they found clams, crabs and an abundance of single cell organisms. The microbes were a source of food for the larger animals. Then began a search for life in the depths, and microbes were found under miles of Arctic ice, and even deep in dry desert sand. Microbes seemed to live everywhere, and many had never seen the sun's light. Hot-water jets deep down were enthusiastically pursued. But there were still huge gaps between simple molecules, larger molecules, and a living cell. Proteins provide a difficult challenge, as they are the workers of the cell and the machines that drive the synthesis of themselves and most other molecules.

A hopeful result involved rocks, particularly minerals. They provided an attractive surface for the assembly of larger molecules. Clays, which have the ability to absorb organic molecules on to their surface, seemed particularly interesting. Indeed it was found that amino acids formed small proteins on clay surfaces when the water evaporated. Other minerals could support the synthesis of somewhat larger protein chains. But this is still a long way from protein synthesis as coded for by nucleic acids.

The cell required a protective membrane that would keep the contents from escaping into the surrounding water. A membrane with lipid molecules was, as we know, the answer. Lipids do not mix with water, but it was found in the 1960s that those from egg yolk mixed with water spontaneously gave rise to tiny spheres – vesicles – with lipid surrounding them. The membrane was formed from two layers of lipid which had their backs away from the water and that part of the molecule that tolerated water in contact with it. This was a nice example of self-organisation of a cell-like structure. Moreover, when RNA was added together with the mixture that had formed small proteins on clays, this resulted in clay particles covered in RNA being found inside the vesicles. Not a cell, but encouraging. There were even theories that clay particles themselves could have formed the first life on earth.

The solution to the origin of the first cell requires three processes to be present: growth, reproduction and evolution. Of these, reproduction and self-replication is the most difficult to solve. It is based on two further key processes: making structures using a source of energy, and being able to pass on such structures from one generation to the next – basic genetics. How these two are linked and how they evolved is very difficult to discover.

One can imagine a self-replicating large molecule that in the right environment could have had copies of itself made, as we now know proteins do this for DNA. But the problem is to establish where the building blocks for this molecule come from. Much more promising are self-replicating networks of interacting molecules. If molecules promote the production of other molecules, one gets closer to a primitive cell. It also comes closer to the chemical reactions which provide the energy for the molecular interactions – neither maintaining nor increasing complexity comes without energy being supplied.

Self-replicating molecules, like DNA, that control other processes in the cell are fundamental for evolution as they are the basis of cell reproduction: their replication passes on to its offspring its own characteristics. In the 1980s the idea that this could have been based on RNA was most exciting. This was due to the discovery of what are known as ribozymes – strands of RNA that are not only carriers of genetic information but also act as catalysts for the synthesis of RNA itself. This led to the idea of an early RNA world.

In this world, it was RNA that carried genetic information and RNA replication was possible in a manner similar to the replication of DNA. Francis Crick speculated that the very first enzyme might have been an RNA molecule that could catalyse the replication of other RNA molecules. There is no such enzyme in current cells but some with similar properties have evolved in the test tube. Starting with a large population of RNAs with random sequences, some of the properties of an RNA enzyme for RNA replication have been formed. Attractive though it is, how it could have arisen is still a mystery. How did that self-replicating RNA come into being in the first place?

One can get a feel for the problem by looking at the chemical structure of RNA. RNA is made of a string of nucleotides each of which contains nine or ten carbon atoms, numerous nitrogen and oxygen atoms, and a phosphate, all linked together in a quite complex structure. There are innumerable different ways of joining up these atoms to form structures not found in RNA, and the formation of the relevant nucleotide remains an improbable event. The Nobel Laureate Christian de Duve has argued for 'a rejection of improbabilities so incommensurably high that they can only be called miracles, phenomena that fall outside the scope of scientific inquiry'. And there remains the problem of the origin of DNA.

The first cell had to divide in order to evolve. This required that the genetic information in the form of nucleic acids had to become enclosed in a membrane, making a primitive cell which could then divide. Again, we just do not know how this happened, but some clues come from studies that suggest that there are interactions between nucleic acids and membranes that could be very important. Experiment showed that membrane-bounded vesicles form when clays are present and encourage lipid molecules to form membranes. These vesicles undergo multiple cycles of growth and division. If a vesicle also contains some material like nucleic acid, then water comes pouring in because of osmosis, and to avoid bursting the vesicles can fuse with empty vesicles and so increase their surface area, and then divide into smaller vesicles. This could, in principle, evolve even further.

The origin of the cell must have been based on Darwinian evolution. A few types of molecule became key players, while others were unstable or were just washed away. They had to have the ability to withstand the sun's rays and cycles of wetting and drying. All these abilities and more refined the chemical mix that eventually gave rise to our ancestor, the first cell, with its self-replicating cycle of molecules. In addition, the cycle had to be variable so that evolution could work by selecting those cycles that survived best.

A different approach to the problem of the origin of life is not to focus on self-replicating molecules, but on small molecules interacting in a network. There are a number of essential requirements for such a system. There needs to be a boundary between the interacting molecules and those not involved, and as we know, cells are enclosed in membranes. But in those ancient times the boundary may have been supplied by membranes made of iron or rock surfaces. There was need for an energy-generating mechanism that involved the transfer of electrons from a rich source to a poor source. Sharp temperature difference or radioactivity could provide this. This energy-giving process had to be linked to keeping the chemical reactions going. The system needed to be able to both evolve and grow. But even with this approach, the origin of life remains a mystery – a mystery that will eventually be solved.

There are those, of course, who claim that there is no real problem or mystery, as there was a designer who created life. For some this is the God of the Bible. There is no evidence whatsoever for such a creator or designer, but those who believe that there is one have no need to think hard about the origin of life. Their views are based on faith and have nothing to do with science. They have

215

the advantage of not having to struggle with the difficult problem of how the cell evolved.

What surprises may there still be, apart from how the first cell evolved? There are still many problems in cell and developmental biology, and the mechanisms whereby cells give the brain its functions is a truly great problem. In very ambitious terms, it may be necessary to understand the function of every protein in a cell. This may require putting all the cell's components on to a computer and thus being able to see in detail what is happening. It may require many years' work, but we would then have a deeper understanding of cells. We would see the complicated chemistry that make cells function.

Even though our cells' origins remain uncertain, we can be certain that whatever we do, or think, or feel, is determined entirely by our cells. We should try to understand them, and protect them. We should always remember that however clever we think cells are, they are cleverer, and that there are still many more surprises to come. Realising that we are the result of the evolution of a cell society, we must respect our cells and look after them as they look after us.

Glossary

Actin a protein molecule involved in muscle contraction and cell movement

Adenosinetriphophate the source of energy for most cellular activities

Amino acids the molecules which are joined together to form proteins

Antigen foreign protein or sugar recognised by the immune system

Apoptosis cell suicide

Asters the star-like structures made of microtubules at the ends of the spindle during mitosis

ATP adenosinetriphophate

Axons the long extensions from nerve cells that carry the electrical signal – the nerve impulse – to the cell's target

Bacteria small cells that have neither a nucleus nor mitochondria

Cell cycle the growth and division of a cell into two cells

Cilia tail-like projections from the cell which beat in a single direction

Chromosomes the string-like package of DNA containing the genes and protein in the nucleus; they become condensed into rod-like structures at cell division

Cloning the process of deriving a number of identical cells from a single cell, such as an egg with a new nucleus inserted

Cyclins molecules that control the cell cycle

Cytoplasm the watery material in the cell outside the nucleus

Dendrites extensions from nerve cells with which other nerve cells form connections (synapses)

DNA the genetic material – a double helix of two chains of nucleotides

Endoplasmic reticulum a set of membranes in the cell where proteins are synthesised

Enzymes protein catalysts that bring about chemical reactions

Filaments proteins joined together in a fine string-like form

Fibroblasts cells of the tissues between structures like bones and muscles and skin

Gastrulation the process in the early embryo in which cell layers such as those that will give rise to the nervous system, the muscles and bone, and the gut, move to their correct position

Gene a region of the DNA that gives the code for the sequence of amino acids in a protein

Genome the heredity information encoded in the chromosomes

Golgi a set of membranes in the cytoplasm which can direct proteins to their location

Ion an atom or a molecule with an electric charge

Lipid fatty molecules

Lymphocytes white blood cells of two types in the vertebrate immune system; type B makes antibodies and type T can kill antigen-presenting cells

Macrophage a type of white blood cell that ingests foreign material

Meiosis the process in the development of germ cells like eggs and sperm when the number of chromosomes is reduced by half

Membrane made of lipid and protein, the membrane separates the cell from the outside world and controls what can come in or out. There are also membranes inside the cell, such as the one surrounding the nucleus

Mesoderm the layer in the early embryo that will give rise to the muscles and skeleton

Metastasis invasion and spread of cancer cells

Micro-RNAs small RNA molecules that can control many processes

Microtubule protein units joined together to form a thin tube in the cell

Mitochondria small sausage-shaped structures in the cell that provide it with energy

Mitosis the stage in the division of the cell when the chromosomes are moved to the daughter cells and the cell is constricted into two cells

Molecules composed of atoms, these are the building blocks of all the components of the cells

Mutation a change in the DNA of the chromosome that alters the coding for a protein or its control region

Myosin a protein that drives contraction by sliding over actin

Nucleotides the small molecules which are joined in a string in nucleic acids like DNA and RNA

Phagocytosis the engulfing by a cell like a macrophage of a bacterium or waste products

Platelets irregularly-shaped, colourless bodies which are present in blood and can form clots to stop bleeding

Protein the key workers in the cell, made of a string of amino acids

Ribosomes small particles in the cytoplasm which construct proteins from amino acid using the code from messenger RNA

Glossary

RNA a string of nucleotides mainly used to take the code for a protein from the nucleus to the cytoplasm

Somites a series of small blocks of tissue along the main axis of the embryo that will give rise to vertebrae

Spindle the structure at mitosis made of microtubules that causes separation of the chromosomes to the daughter cells

Stem cells cells that divide repeatedly, one of the daughter cells developing into a different type of cell while the other remains a stem cell

Synapse the junction between an axon and another nerve cell

Telomere the ends of the chromosomes that are reduced in length each time the cell divides

Thrombus a blockage in a blood vessel

Transcription the transfer of the code for a protein from DNA in the chromosome to messenger RNA

Vesicles small membrane-bound spheres with a hollow interior

Virus a tiny infectious agent that can only reproduce inside cells

References

Alberts, B., et al (2006), *Essential Cell Biology.*
Garland, 2nd edn.

Bowler, P. (1989), *The Mendelian Revolution.* Athlone
Press.

Finkel, T., et al (2007), 'The common biology of cancer
and ageing'. *Nature*, 448, 767–74.

Harris, H. (1999), *The Birth of the Cell.* Yale University
Press.

Dreamers, D. W, and Hazen, R. M. (2005), *Genesis:
The Scientific Quest for Life's Origins.* Henry Press.

Kandel, E. R., et al (2000), *Principles of Neural Science.*
McGraw Hill, 4th edn.

Kirkwood, T. B. L. (2005), *Ageing: Understanding the
Odd Science of Ageing Cells.*

Rando, T. A. (2006), 'Stem cells, ageing and the quest
for immortality'. *Nature*, 441, 1080–6.

Shapiro, R. (2007), 'A simpler origin of life'. *Scientific
American*, 296, 25–31.

Wolpert, L., et al (2007), *Principles of Development.*
OUP, 3rd edn.

Index